Topics in Heterocyclic Chemistry

Volume 61

The series Topics in Heterocyclic Chemistry presents critical reviews on present and future trends in the research of heterocyclic compounds. Overall the scope is to cover topics dealing with all areas within heterocyclic chemistry, both experimental and theoretical, of interest to the general heterocyclic chemistry community. The series consists of topic related volumes edited by renowned editors with contributions of experts in the field. All chapters from Topics in Heterocyclic Chemistry are published OnlineFirst with an individual DOI. In references, Topics in Heterocyclic Chemistry is abbreviated as Top Heterocycl Chem and cited as a journal.

Eric Nacsa
Editor

Construction and Functionalization of Heterocycles by Electrochemistry

With contributions by

P. Fang · J.-F. Guo · X. Hu · L. Lin · T.-S. Mei · T. Prenzel · Z. Ruan · A. Sandvoß · Z. Shi · S. R. Waldvogel · J. Wang · J. Wen · J. Winter · J. Wu · J. Yang · K.-Y. Ye · Y. Yin · Y. Zhao · R. Zhu

Springer

Editor
Eric Nacsa
Department of Chemistry
The Pennsylvania State University
University Park, PA, USA

ISSN 1861-9282 ISSN 1861-9290 (electronic)
Topics in Heterocyclic Chemistry
ISBN 978-3-031-74263-7 ISBN 978-3-031-74264-4 (eBook)
https://doi.org/10.1007/978-3-031-74264-4

This Springer imprint is published by the registered company Springer Nature Switzerland AG
The registered company address is: Gewerbestrasse 11, 6330 Cham, Switzerland

Preface

It is hard to overstate the importance of heterocycles as valuable chemical substructures. In biologically active molecules, their polarity, hydrophilicity, and H-bonding capabilities play important roles and account for their prominence in pharmaceuticals and agrochemicals. Their unique electronic structures also make them versatile units in ligands, conjugated materials, and many other useful compound classes. Although they are often relatively small molecules, heterocycles represent an incredibly diverse set of structures. Efficient access to their full range of product architectures thus calls for a similarly diverse set of synthetic approaches.

The constant demand for new products, combined with growing global population, makes the need for sustainable synthesis increasingly pressing. As a synthetic platform, electrochemistry meets the needs of modern methods in many ways, which explains its recent re-emergence. By selectively engaging most redox-active functionalities, it can generate highly reactive intermediates under mild conditions. This outcome enables the conversion of simple and abundant compounds into complex, valuable products in shorter sequences than may be possible by other means. It also offers unique opportunities for improving atom economy. In particular, oxidative transformations are readily accomplished without chemical oxidants owing to efficient cathodic hydrogen evolution.

Several leading synthetic organic chemistry groups around the world have been leveraging electrochemistry in innovative ways over the past decade. In the context of preparing heterocyclic products, both the construction and functionalization of several classes have received significant attention. This volume includes a range of mechanistic roles for electrochemical activation, from the generation of radicals and the redox modulation of metal catalysts for heterocycle synthesis to the reduction and oxidation of heterocyclic cores for their peripheral functionalization. It also surveys a wide range of core architectures that can be accessed and manipulated.

The first chapter focuses on the electrochemical synthesis and derivatization of indoles, one of the most important heterocycles in biological settings. The second chapter covers a wider range of heterocycles, using electrochemistry and metal catalysis to access pyrones and pyridines, in addition to functionalizing substrates such as quinolines and tetrahydroisoquinolines. The third chapter includes oxidative benzylic azolation, the addition of selenium-containing groups to heterocycles, and further contributions to indole synthesis and functionalization. The fourth chapter describes electrochemical radical generation and cycloaddition to forge C–N and N–N bonds, affording a rich suite of products such as pyrazoles, isoxazoles, imidazoles, triazoles, and several others. The fifth chapter focuses on indole dearomatization, which provides access to stereochemically dense products, including spirocycles. The sixth chapter discusses electrochemically generated N-centered radicals for cyclizations, spirocyclizations, and migratory cyclizations, as well as electrochemically generated nitrilium ions as precursors to N-heterocycles. The seventh and final chapter covers topics from the reductive functionalization azines to the oxidative introduction of alkyl and sulfur-containing groups onto heterocycles.

Finally, the authors are thanked extensively for their conscientious efforts preparing these chapters and throughout the editing process. Each chapter reflects the unique contributions and perspectives of its authors, which will hopefully provide variety across this volume and keep readers engaged.

University Park, PA, USA Eric Nacsa

Contents

Top Heterocycl Chem (2024) 61: 1–22
https://doi.org/10.1007/7081_2024_79

Published online: 16 October 2024

Construction and Derivatization of Indole Derivatives Under Electrochemical Conditions

Ling Lin and Pengju Feng

Contents

Abstract Indole and its derivatives, both naturally occurring and synthetic, are known to possess wide-ranging biological activity, which promote chemists to develop numerous methods for their efficient preparation. Organic electrocheical synthesis, possessing unique reaction selectivity, offers a mild and convenient alternative to the synthesis of indole derivatives versus conventional thermal transformation using hazardous chemical oxidants and reductants. This chapter will selectively introduce the examples of indole synthesis from various reactants, indole functionalization from simple indole derivatives, and construction of three-dimensional indole-based molecules via electrochemical protocols.

Keywords Green chemistry · Indole derivatives · Indole functionalization · Organic electrosynthesis · Three-dimensional indole derivatives

Indole and its derivatives, both naturally occurring and synthetic, are known to possess wide-ranging biological activity [1–4]. A number of drug molecules already

L. Lin and P. Feng (✉)
Jinan University, Guangzhou, Guangdong, China
e-mail: pfeng@jnu.edu.cn

on the market contain at least one indole motif, such as vinblastine, tadalafil, rizatriptan, and so on [5]. Thus, the importance of obtaining indole derivatives cannot be overstated. Since indole was firstly prepared by Adolf von Baeyer in 1866 [6], extensive efforts have been devoted to expand the diversity of indole scaffolds [7]. There are already numerous well-known named reactions, such as the Fisher indole synthesis, Bartoli indole synthesis, Reissert indole synthesis, Cadogan-Sundberg indole synthesis, Larock indole synthesis, Gassman indole synthesis, Bischler indole synthesis, Batcho-Leimgruber synthesis, and so on [7–14]. Meanwhile, three-dimensional heterocyclic frameworks are common in intricate architectures with interesting biological properties [15, 16]. The dearomatization of indoles provides a powerful method for generating such structural diversity and molecular complexity. Many elegant methods for dearomative difunctionalization of indoles have been established, including transition metal-catalyzed transformations, cycloadditions, radical-mediated reactions, and electrophile-triggered functionalizations [17–20]. Despite this great progress in the synthesis of indoline-based heterocycles, demand for more economical and environmentally friendly protocols to avoid expensive transition metals, excess external oxidants, or tedious workup procedures still exists.

Recently, organic electrosynthesis has been rediscovered as mild and efficient alternative to traditional thermal chemistry for organic synthesis with electrons as reactants, avoiding using hazardous chemical oxidants and noble metal catalysts [21–28]. Meanwhile, electrochemistry can provide new reaction pathways that are complementary to conventional synthetic methods. In particular, organic electrosynthesis shows great advantages for the synthesis and derivatization of electron-rich heterocycles, which has been constantly investigated with fruitful outcomes. In recent years, numerous elegant and practical electrochemical transformations for the synthesis of indoline-based heterocycles have been developed [29, 30]. In addition, along with the rapid development of electrochemical reaction setup and testing instruments, a better understanding of the electrochemical reaction mechanisms was illustrated in recent discoveries. This chapter will focus on representative electrochemical syntheses and transformations of indoline-based heterocycles. We also hope that readers might find inspiration to apply these methods in new settings or develop new synthetic methodologies.

1 Electrochemical Synthesis of Functional Indole Derivatives

Electrochemical strategies employ traceless electron as redox reagents, facilitating chemical transformations without exogenous chemical oxidants/reductants or noble metal catalysts. They also occupy an increasingly important place in synthesis of indole derivatives via C-N/C-C bond formation, owing in part to their green and sustainable nature [31]. The indole precursors, which are usually inert under traditional thermal chemistry, could be efficiently transformed to highly reactive radical intermediates through direct oxidation/reduction at an electrode or through redox mediator-assisted electrolysis. Structurally versatile indoles with a wide range of functional groups were rapidly generated by a novel reaction pathway. We will present the synthesis of indole derivatives from the following three aspects: (1) electrochemical synthesis of indole derivatives via intramolecular cyclization, (2) electrochemical synthesis of indole derivatives via intermolecular cyclization, and (3) electrochemical synthesis of indole derivatives via indole modification.

1.1 Electrochemical Synthesis of Indole Derivatives via Intramolecular Cyclization

Electrochemical generation of *N*-centered radicals is frequently involved as key step to synthesize indole derivatives [32]. For example, Xu and coworkers developed an elegant electrochemical method initiated by *N*-radicals for the synthesis of indoles and azaindoles employing the inexpensive [Cp_2Fe] as the redox-relay reagent (Scheme 1) [33]. They found that aryl- and heteroarylamines with tethered alkynes (**1**) could undergo efficient tandem C-N and C-C coupling to form two heterocyclic

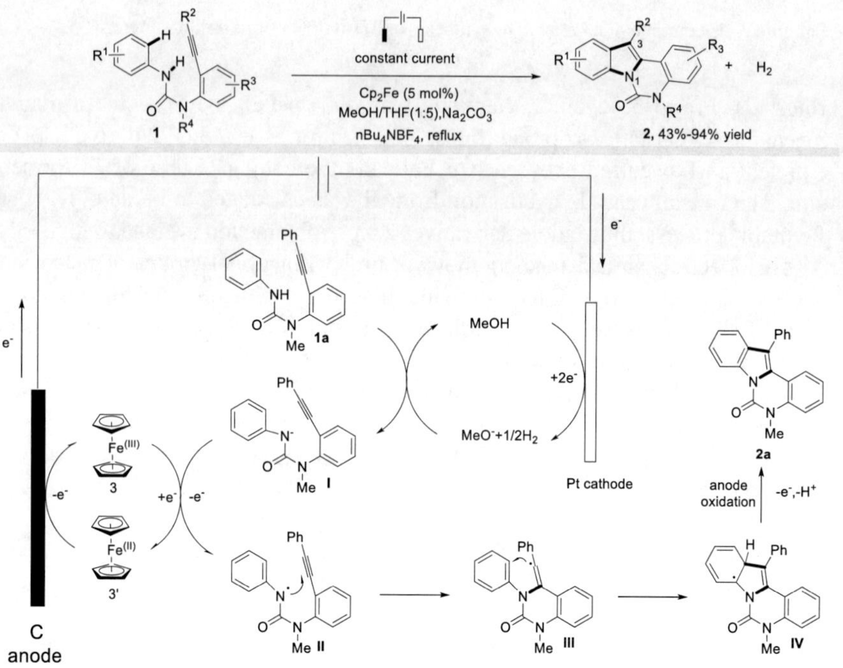

Scheme 1 Electrochemical C-H/N-H functionalization for the intramolecular synthesis of (aza) indoles

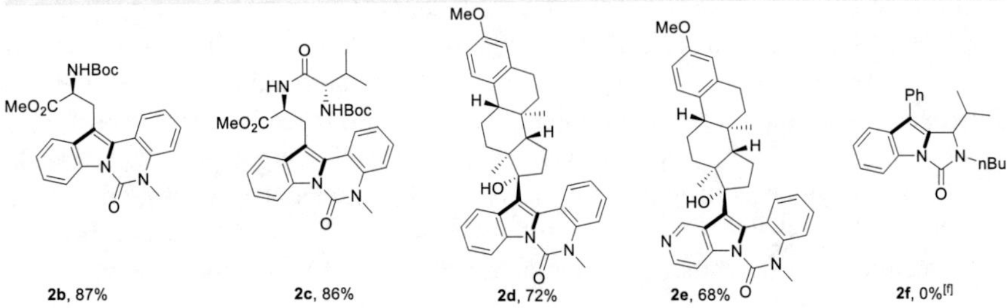

Scheme 2 Selected examples of cyclization products from Scheme 1

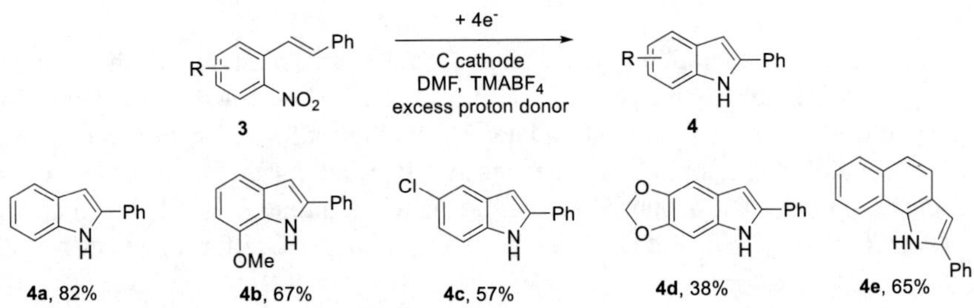

Scheme 3 Electrochemical reductive synthesis of 1*H*-indoles from *o*-nitrostyrenes

rings (**2**). From the cyclic voltammograms of [Cp_2Fe], they provide a plausible mechanism. [Cp_2Fe] was prone to lose one electron at the anode to give [Cp_2Fe]$^+$. Efficient SET occurred between [Cp_2Fe]$^+$ and the conjugate base of **1a** formed in situ. The overall cascade cyclization from **II** was calculated to be an energetically favorable process; thus indole derivatives **2** were obtained in average to high yields.

This electrochemical method shows a broad functional-group tolerance in the successful preparation of indoles with diverse substituents, including halogens, ketone, aldehyde, free alcohol, orthoester, acetal, and so on. Notably, substrates with a chiral Boc-amino ester (**2b**) and dipeptide (**2c**) could smoothly transform to indoles in high yield (Scheme 2). Indoles and azaindoles with a complex estrone branch (**2d** and **2e**) were successfully formed via electrolysis in good yield. Imidazo [1,5-a]indol-3-one derivative (**2f**) with a relatively strained five-membered ring could not form under these electrolysis conditions, however.

Electroreductive strategies also show the possibility to transform *ortho*-nitrostyrenes **3** to 1*H*-indoles **4** via a 4-e$^-$ reduction/cyclization process (Scheme 3). In 2011, Peters and coworkers reported a novel procedure to synthesize 1*H*-indoles **4** based on the direct, room temperature electrochemical reduction using a carbon cathode in DMF [34]. This overall four-electron conversion of the *ortho*-nitrostyrenes could deliver compounds **4a-4e** in moderate to good yields after a relatively short electrolysis (40 min). They found that the identities of different

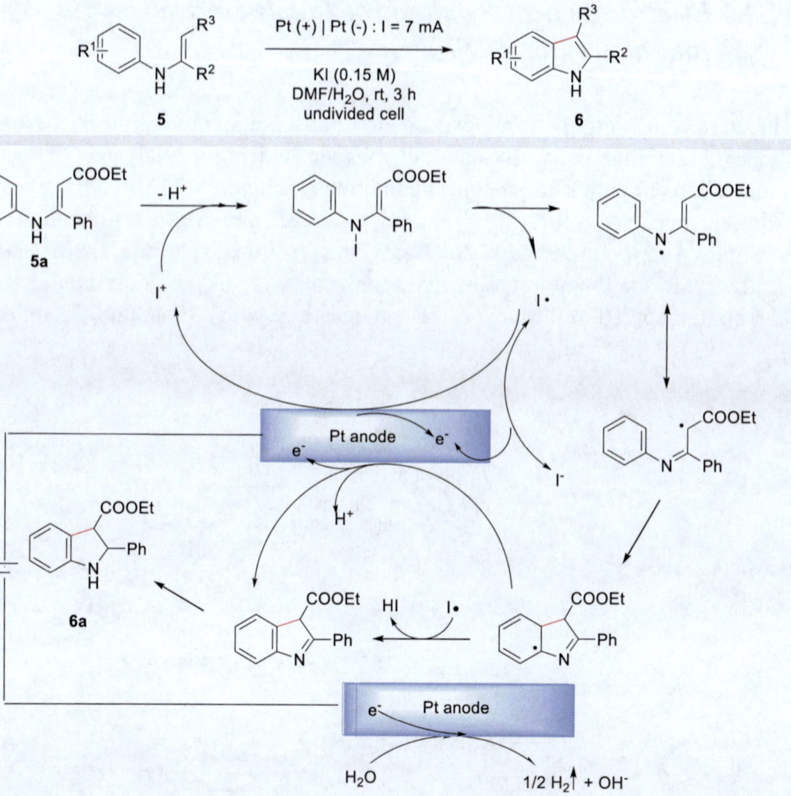

Scheme 4 Iodide-mediated electrochemical synthesis of substituted indoles via intramolecular oxidative annulation

proton donors and solvents affected the efficiency. The substrate compatibility and practicality of this electrolysis reaction remain to be explored.

Lei and coworkers found that KI could act as both electrolyte and redox mediator to promote indole formation from *N*-aryl enamines via electrochemical intramolecular dehydrogenative annulations (Scheme 4) [35]. Various *N*-aryl enamines with different functionalities could efficiently transform to 2,3-subtituted indoles under oxidant-free and transition metal-free conditions in an undivided cell. Moreover, when *N*-pyridyl enamines were applied as the substrates, imidazo[1,2-a]-pyridines could be obtained in high yield. Mechanistic studies revealed that the transformation was initiated by an iodonium intermediate (I^+), generated from iodide ion via two-electron anodic oxidation, and reaction with *N*-aryl enamine **5a** to form an *N*-iodo intermediate. A cyclization process was followed by the homolysis of the N-I bond. The desired product was formed after intramolecular radical addition, anodic oxidation, and deprotonation.

1.2 Electrochemical Synthesis of Indole Derivatives via Intermolecular Cyclization

In 2018, Xu's group developed an intermolecular indole synthesis using easily accessible aniline derivatives and alkynes via ruthenium-catalyzed C-H activation/annulations under electrochemical conditions (Scheme 5) [36]. They found that the dimeric precatalyst $[RuCl_2(p\text{-cymene})]_2$ could be converted to a ruthenium diacetate complex **I** in the presence of NaOAc. Complex **I** then follows a multistep reaction pathway: formation of unproductive complexes with alkyne or productive formation of six-membered ruthenacycle **II** via reacting with compound **7**, followed by

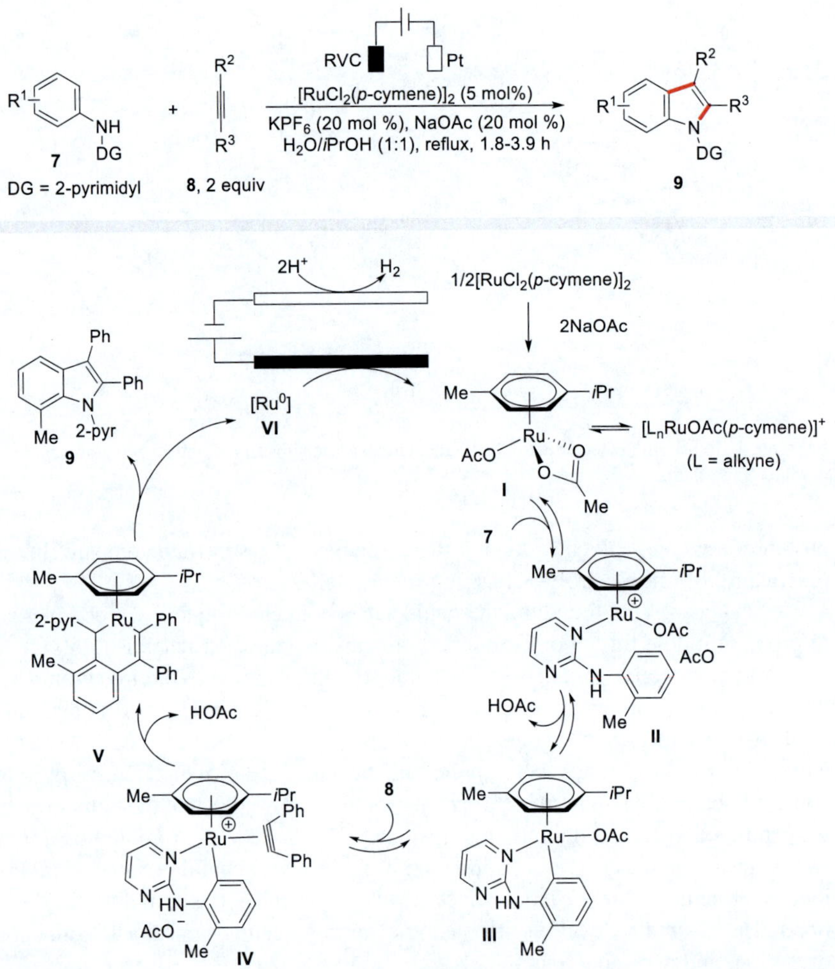

Scheme 5 Indole synthesis via ruthenium-catalyzed electrochemical dehydrogenative alkyne annulation

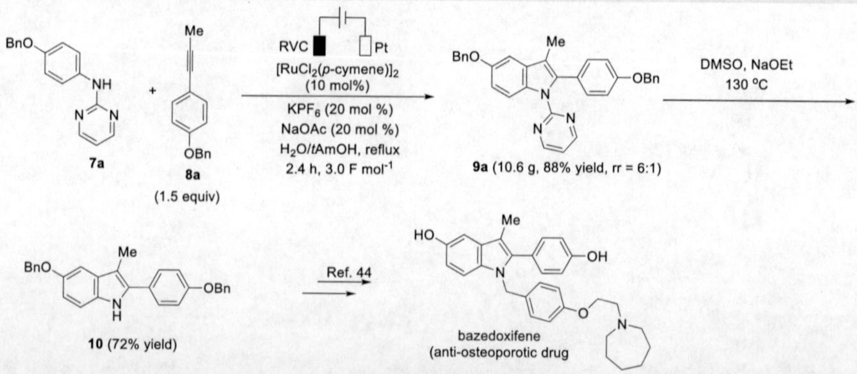

Scheme 6 A synthetic application of electrochemical intermolecular cyclization (Scheme 5)

reversible C-H activation. The acetate ligand of **II** could be subsequently displaced by alkyne **8** as a result of π-coordination to give intermediate **IV**. The six-membered ruthenacycle **V** could be obtained by migratory insertion of the alkyne into the Ru-C bond, following the dissociation of the pyrimidyl nitrogen from the metal center. The desired indole product **9** and Ru(0) species (**VI**) were ultimately delivered via reductive elimination of **V**. Catalytically active Ru^{II} complex **I** could be regenerated by anodic oxidation of **VI**. The transformation was carried out in an undivided cell wherein protons are reduced at the cathode to generate H_2, obviating the need for H-atom acceptors.

The synthetic utility of this electrochemical dehydrogenative alkyne cyclization was demonstrated by synthesis of a multisubstituted free indole **10**, a critical precursor to synthesize the antiosteoporotic drug bazedoxifene (Scheme 6) [37]. Indole **9a** was easily prepared on 10-g scale in 88% yield via the electrochemical annulation of aniline derivative **7a** and alkyne derivative **8a**. The pyrimidyl directing group was then removed by a facile established method under basic conditions to obtain free indole **10** in good yield.

1.3 Electrochemical Synthesis of Indole Derivatives via Indole Modification

Among the many atom-economical C-H/N-H cross-coupling strategies for C-N bond construction, methods for direct C-H amination of heteroarenes have received tremendous attention [4, 38, 39]. Feng and coworkers envisioned that the azolation of five-membered indole ring could be achieved via an electrooxidative heteroarene amination pathway through tuning the electron density of indoles with different *N*-protecting groups (Scheme 7) [40]. Unlike the electrochemical oxidative dehydrogenative C-H amination protocol with *N*-alkyl-protected indoles and *N*-alkylsulfonamides developed by Ackermann and coworkers, Feng used a constant

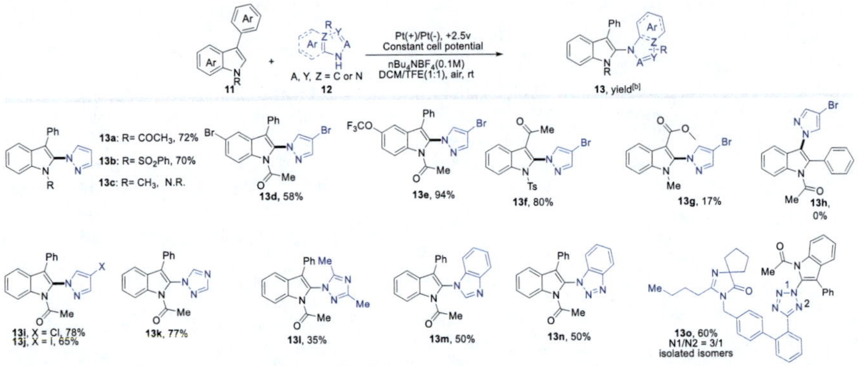

Scheme 7 Selected examples of electrooxidative azolation of indoles

cell potential electrolysis at room temperature to couple *N*-acyl-protected indoles with various azoles in good efficiency. They found that the protecting group of indole was crucial to the success of azolation. Replacement of the acyl group with a benzenesulfonyl group gave **13b** in similar 70% yield. However, no product (**13c**) was formed when *N*-methyl-3-phenyl indole was used in the azolation. Coupling 3-bromopyrazole with various *N*-acyl-protected C3-substituted indole derivatives under standard electrochemical conditions gave the desired products in good yields (**13d-13g**), while the 2-phenyl indole derivatives were prone to overoxidation to give a black mixture without generating the desired product (**13h**). A wide range of functional pyrazoles, benzimidazole, triazole, benzotriazole, and tetrazole were found to be fully tolerated by the optimized electrooxidation (**13i-13n**). Particularly, the commercially available drug Irbesartan with a tetrazole tail was smoothly transformed by the electrochemical azolation process to give product **13o** in 60% yield [41].

In addition, they extended the electrochemical oxidative C-H/N-H cross coupling for the synthesis of C2/C3 azolated indole/thiophene with hydrogen evolution to 5-membered heteroarenes with a halogen substitution on the C2/C3 position. It is well known that halogen substituents are often destroyed in metal-catalyzed processes. In addition, 2- or 3-haloindoles prepared by C-N bond formation could undergo accompanying HX elimination under simple thermal heating condition [42–45]. Thus, the direct C-N bond formation between azoles and the five-membered heterocycles with halogen substitution on the C2/C3 position presents a clear synthetic challenge. Initially, they achieved the electrochemical transformation with Ts-protected 3-bromoindole as model substrate to couple with simple pyrazole. The reaction was conducted at a constant current of 10 mA in a mixed solvent (DCM/TFE = 1:1 v/v) using n-Bu_4NBF_4 (0.1 M) as supporting electrolyte, a user-friendly undivided cell setup, and with a platinum plate cathode and anode at 50°C under air.

A variety of halo-indoles/thiophenes proved suitable as substrates, providing C2/C3-azolated halo-indoles/thiophenes in good to excellent yields (Scheme 8). For

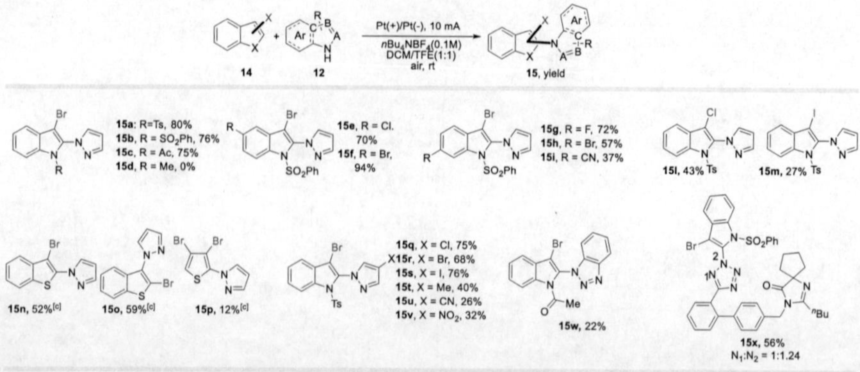

Scheme 8 Selected examples of electrooxidative azolation of 2- and 3-haloindoles

indole substrates, the *N*-protecting group is important for the success of this transformation [46]. Electron-deficient protecting groups that can be easily removed under basic condition after azolation could deliver the desired product in 75–80% yield (**15a-15c**), while one with an *N*-alkyl group that is prone to overoxidation gave an unidentified mixture (**15d**). Various *N*-benzenesulfonyl bromoindole derivatives with a wide range of substituents on the phenyl ring were well tolerated by this transformation (**15e-15i**). C3 chloro and iodo indole were also compatible with the reaction parameters, delivering the corresponding C-H/N-H coupling products in 43% and 27% yield, respectively (**15l-15m**). The direct C-N bond formation reaction could extend to benzothiophene with a bromide at the C2 or C3 position, generating the desired product in comparable moderate yields (**15n**, 52%; **15o**, 59%). Various azoles with a wide range of substituents were also suitable substrates for coupling with bromo indole/benzothiophene derivatives under constant current electrolysis (**15p-15w**). Electron-deficient pyrazoles usually react inefficiently (**15v**). Notably, this method allows chemists to reliably perform a late-stage azolation to access complex C3 bromo indole **15x** in 56% yield, the bromo substituent of which could serve as a handle for future late-stage modifications. Their preliminary mechanistic study revealed that the reaction proceeds through an intermolecular radical combination between the heteroarenes and azoles.

Apart from azoles, easily accessible aniline derivatives represent versatile motifs for C–N bond formation; thus explorations of their electrochemical transformations are synthetically important. However, due to homocoupling of *N*-centered radicals derived from anilines and product overoxidation, synthetically useful examples of intermolecular aryl C–H amination with anilines using electrochemistry are rare [47–49]. In 2021, Feng and coworkers developed a mild and general protocol for effective amination of five-membered heteroarenes with aniline derivatives under electrochemical conditions [50]. The generality of this cross-coupling reaction was proved by a wide range of indoles and aniline derivatives (Scheme 9). *p*-Methoxy aniline with different protecting groups, such as Ts-, α-naphthSO_2-, 2-thiophenSO_2-, -CO_2Et, and –$CO_2{}^t$Bu, could efficiently react with an Ac-protected indole derivative

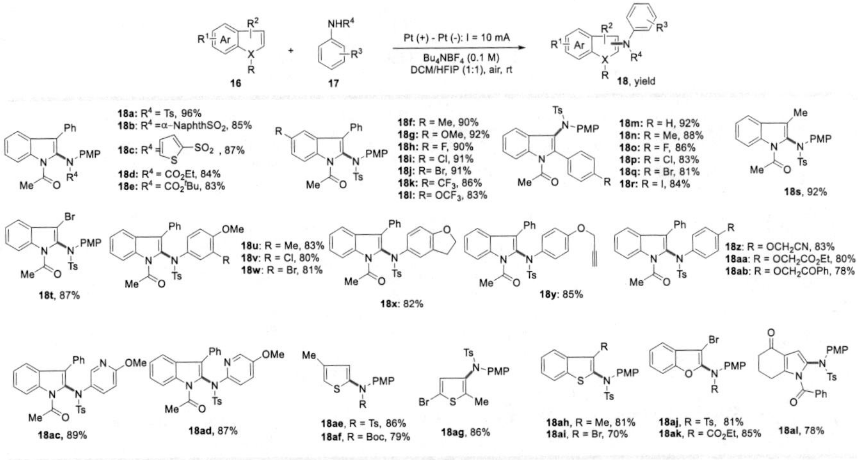

Scheme 9 Selected examples of electrooxidative C-H amination of heteroarenes with aniline

to deliver the desired products in high yields (**18a-18e**). Aryl rings with different functionalities on the 2- or 3-position of indole were compatible with this electrochemical coupling reaction, with accompanying high isolated yields (**18f-18r**). The selective electrochemical C-H amination proceeded smoothly with a 3-methyl indole (**18s**) or even a 3-bromide indole (**18t**) as substrates without touching the methyl and bromo groups. The electrochemical amination also demonstrated good tolerance of Ts-anilines with various substituents (**18u-18ab**), including halogen, cyano, ketone, ester, heterocycles, and alkyne. Pyridine derivatives were also good candidates to participate this electrochemical transformation, delivering products in high yields (**18ac-18ad**). In addition, the electrolysis reaction was successfully extended to other five-membered heteroarenes with *para*-methoxy aniline as the coupling partner under optimal electrochemical conditions (**18ae-18al**).

To demonstrate the practicality of above electrochemical reaction, they tried different ways to further modify the product. 2-Amino indole **19** could be obtained in 93% yield by simply removal of Boc group of **18e** with TFA, which could be further transformed to amide **20** and dihydroindolo[2,3-b]indole **21** by oxidative removal of PMP and a palladium-catalyzed oxidative coupling reaction, respectively (Scheme 10a) [51, 52]. For mechanistic investigations, they found that homocoupling products **22** and **23** were isolated in different yields when conducting the reactions using some specific substrates (Scheme 10b) [53]. These compounds may be formed through nitrogen-radical recombination and indole radical cation recombination, respectively. These results suggested that nitrogen-centered radical derived from the aniline and the indole radical cation simultaneously formed during the electrolysis. The desired product was formed by radical cross-coupling between these *N*- and *C*-centered radicals.

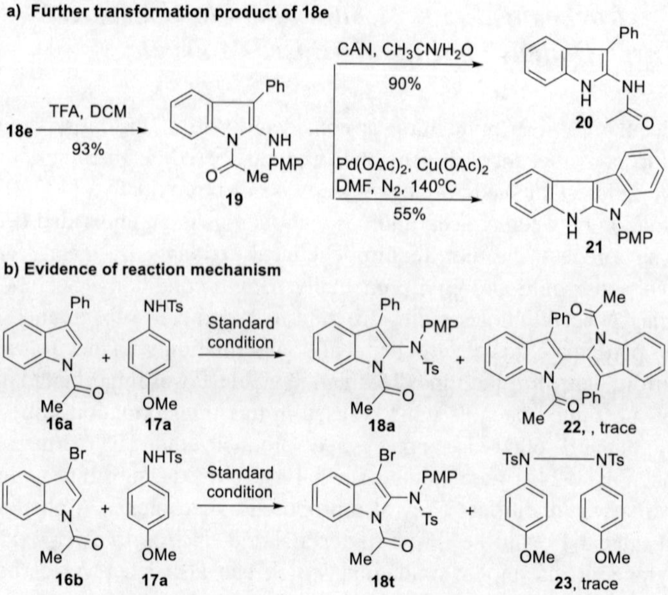

Scheme 10 Product transformations and mechanistic studies from Scheme 9

2 Electrochemical Dearomatization of Indole Derivatives

Dearomatization of indoles has been recognized as a powerful method for generating structural diversity and complexity of three-dimensional indole-based molecules and received considerable attention [54–57]. Indeed, many elegant methods have been established to showcase the construction of indoline/indolenine-based alkaloids from planar indoles [58–61]. Apart from the classic dearomatization of indoles that take advantage of the inherent nucleophilicity of C3 position [62, 63], the development of oxidation-induced C-H functionalization paves the way to use umpolung of indoles for opening up another side of dearomative indole chemistry [64]. The common disadvantage of recent oxidative dearomatization lies in their requirement for external stoichiometric oxidants and/or expensive catalysts. Recently, electrochemistry has been rediscovered as mild and efficient alternative to traditional thermal chemistry for organic synthesis with electrons as reactants, avoiding using hazardous chemical oxidants and noble metal catalysts [65]. Meanwhile, electrochemistry can provide new reaction pathways that complement conventional synthetic methods. In this regard, several elegant strategies for oxidative dearomatization indoles have been developed. We will present the electrochemical **dearomatization of indole derivatives** from the following two aspects: (1) electrochemical [3 + 2] annulation to construct three-dimensional indole-based molecules and (2) 2,3-difuctionalization of indoles via electrochemical dearomatization.

2.1 *Electrochemical [3 + 2] Annulation to Construct Three-Dimensional Indole-Based Molecules*

Intermolecular [3 + 2] annulation is considered one of the most straightforward methods to construct three-dimensional five-membered heterocycles. In 2017, Lei and coworkers established an elegant stepwise electrooxidative [3 + 2] annulation transformation between *N*-acetylindoles and phenols in an undivided cell [66]. The electrolysis process did not require chemical oxidants or metal catalysts and proceeded under mild and environmentally benign conditions (Scheme 11). They found that *N*-acetylindoles with different C2 and C3 substituents (**25**) could smoothly participate this electrochemical transformation with electron-rich phenols (**24**) to afford benzofuroindoline (**26**) in good yields. It was important to mention that methoxy-substituted phenols must be used in this transformation, especially *para*-methoxy phenol; otherwise no desired product could be formed. These *p*-alkoxyphenols (**24a**) are thus special, activated substrates under these electrooxidative conditions. They also proposed a mechanism. A phenolic oxygen-centered radical **I** could be formed after a single-electron-transfer oxidation of *p*-methoxylphenol via anodic oxidation, which can also react via carbon-centered radical resonance contributor **II**. In the meantime, indole radical cation **III** was

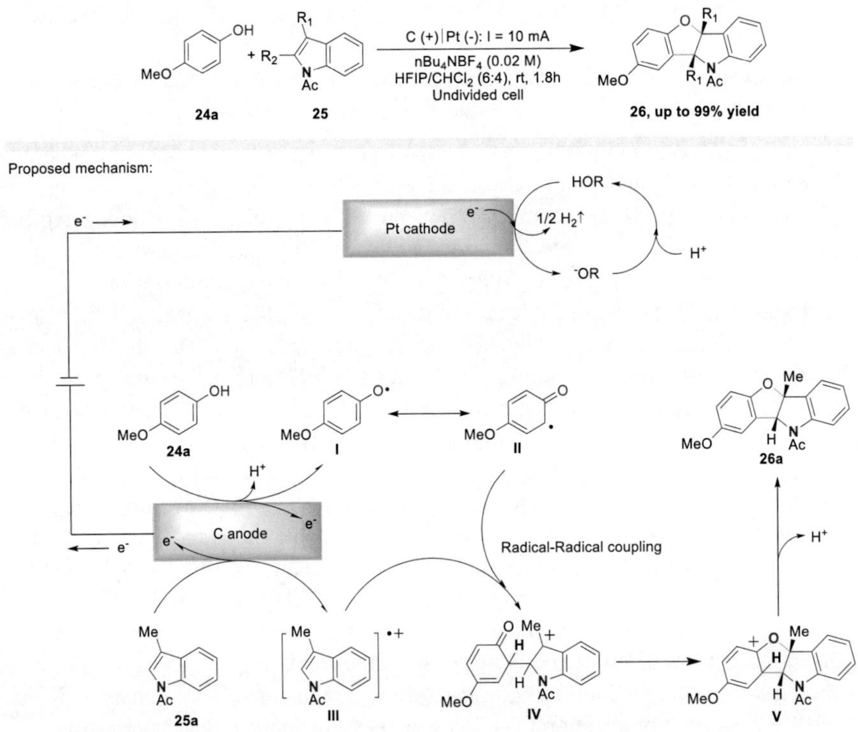

Scheme 11 Electrooxidative [3 + 2] annulation between phenols and indoles

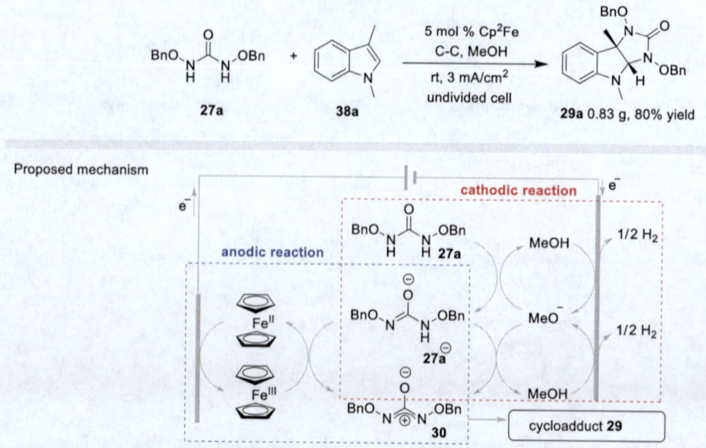

Scheme 12 Electrooxidative generation of diaza-oxy-allyl cation and its application to cycloadditions

afforded by anodic oxidation. The radical-radical coupling between **II** and **III** gave cation intermediate **IV**, which subsequently proceed intramolecular cyclization and deprotonation for generation-desired product benzofuro[3,2-b]indoline **26**.

The diaza-oxyallyl cation was considered a useful reactive intermediate for cycloaddition with carbon-carbon double bonds or dienes to provide biologically important diamine motifs [67, 68]. In 2018, Luo and coworkers developed a straightforward electrochemical process under oxidant-free, base-free, and electrolyte-free conditions with Cp_2Fe as redox mediator to generate diaza-oxylallyl cations and successfully applied it to synthesize diamine motifs with various *N*-alkyl protected indoles (Scheme 12) [69]. They found that the reaction could not proceed in a divided cell. That means the base generated in situ at the cathode is very important to this [3 + 2] cycloaddition reaction. They proposed a mechanism wherein Cp_2Fe could undergo anodic oxidation to $[Cp_2Fe]^+$ and solvent MeOH was reduced to deliver methoxide anion and hydrogen gas at the cathode. The two in situ generated active species then worked as oxidant and base to facilitate the oxidation of urea **27a** to diaza-oxyallyl cation **30**, which could readily participate [3 + 2] or [4 + 3] cycloaddition with double bonds and dienes to give the desired products **29**.

2.2 *2,3-Difunctionalization of Indoles via Electrochemical Dearomatization*

Facile electrochemical difunctionalization of the double bond embedded in the indole heterocycle provides a protocol to gain three-dimensional, value-added heterocycles. In 2019, Vincent and coworkers developed an electrochemical oxidative

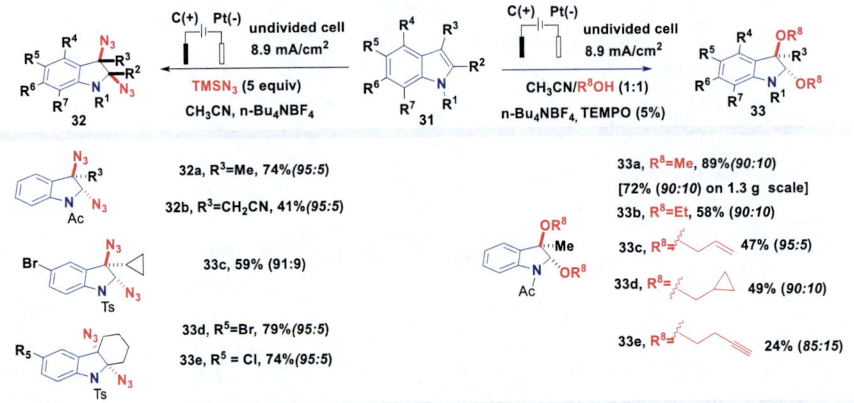

Scheme 13 Selected examples of electrochemically 2,3-difuctionalized indoles

dearomatization of indoles offering 2,3-diazido and 2,3-dialkoxy indolines at a constant current in an undivided cell (Scheme 13) [70]. Both azide and alcohols were used as nucleophiles in this electrochemical difunctionalization (**32a-32e** and **33a-33e**). This reaction showed good functional-group tolerance, accommodating cyano, halogen, alkyne, alkene, and cyclopropane groups. Both 3-substituted and 2,3-disubstituted indoles showed good reactivity under the same oxidative conditions. It was important to mention that the indoles should have an electron-deficient *N*-protecting group; otherwise no desired product was detected.

The reaction mechanism of this difunctionalization that was proposed involved *N*-substituted indole **31** first being oxidized at the anode to give a radical cation intermediate **I**, which underwent an electrophilic addition reaction at C3-position to methoxide or azide anions formed directly and indirectly at the cathode (Scheme 14). This addition process could subsequently deliver a radical intermediate at the C2 position of indole, which was readily oxidized to give iminium ion **II**. The methoxide or azide then reacted again as nucleophiles to attack iminium ion **II**, eventually delivering **32** or **33** as *cis*- or *trans*-isomers.

Notably, electrochemistry can provide new selectivities that are totally different from those obtained in well-documented approaches. Feng and coworkers achieved a novel electrooxidative dearomatization strategy for synthesizing 2-oxindole motifs. The reaction could be simply tuned to form 3-hydroxy-2-oxindole derivatives or 3,3-diaryl-2-oxindole derivatives by merely altering anodic materials under the same reaction conditions [71]. Firstly, using graphite sheet as anode, 3-hydroxy-2-oxindole derivatives could be efficiently synthesized (Scheme 15). The *N*-Ac-3-phenyl indole substrate gave 1-acetyl-2-oxo-3-phenylindolin-3-yl acetate (**35a**) in 85% yield under the optimal electrochemical conditions. Various functional groups on the benzene ring of indole resulted in slightly decreased yield of products (**35b-35h**). Product **35i** was obtained in 70% yield via electrochemical dearomatization of 1-(3-phenyl-1*H*-benzo[f]indol-1-yl)ethan-1-one, whose structure was further confirmed by X-ray single crystal diffraction. *N*-Ac-protected indole with various aryl

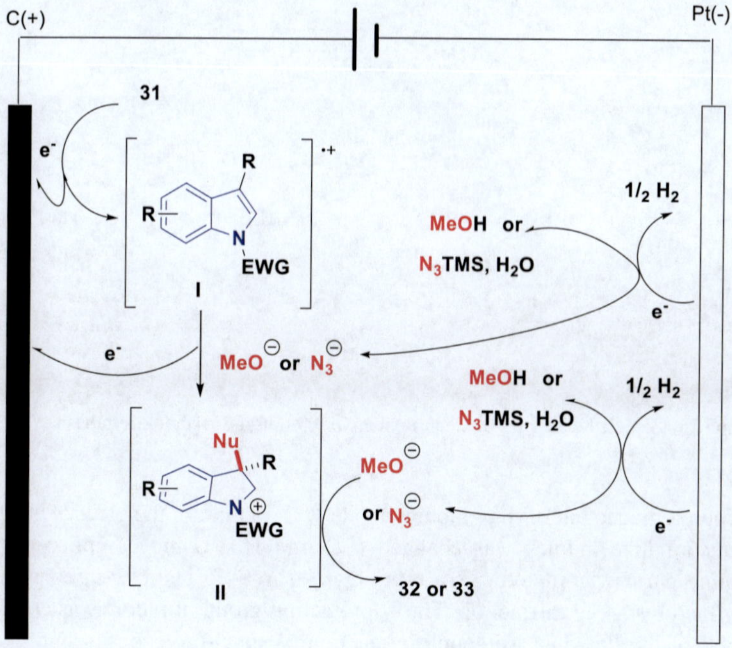

Scheme 14 Proposed reaction mechanism for electrochemical difunctionalization of indoles (Scheme 13)

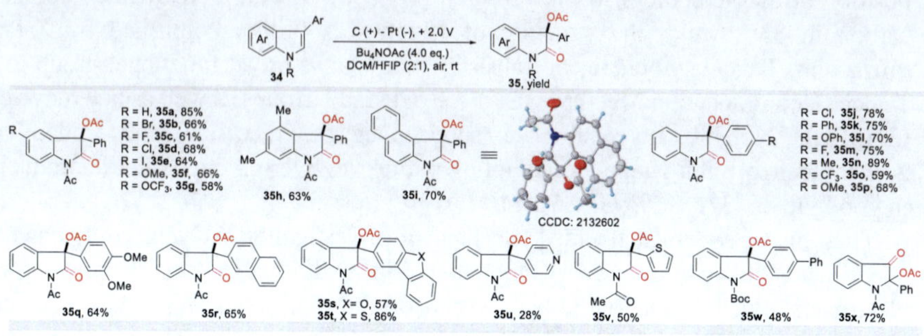

Scheme 15 Selected examples for electrooxidative synthesis of 3-aryl-3-hydroxyl-2-oxindole derivatives

rings at the C3 position were good substrates for this electrochemical transformation, giving the desired products in 59–89% yield (**35j-35t**). When heterocyclic rings were located on the C3 position of indole, the electrochemical transformation delivered the products in low yields, as shown for pyridine (**35u**) and thiophene (**35v**). Replacement of Ac with other electron-withdrawing protecting groups for indoles could also deliver the product in acceptable yield (**35w**, 48%). Notably, this tactic was also suitable for synthesis 3-oxoindoline (**35x**, 72%).

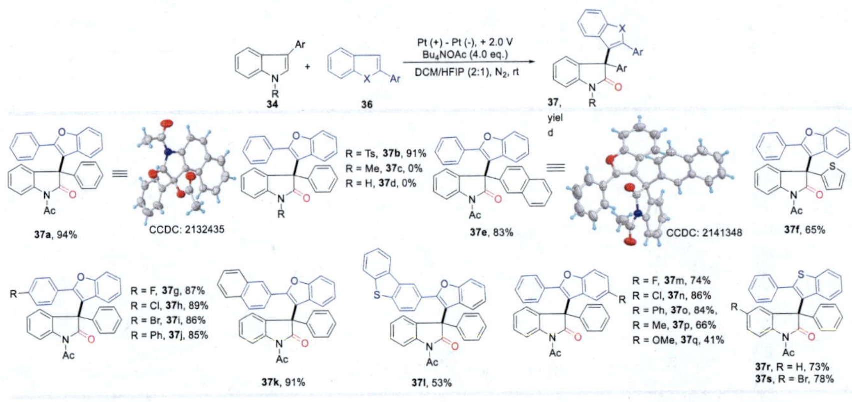

Scheme 16 Selected examples of electrooxidative 3,3-diaryl-2-oxindole synthesis

Upon replacement of the graphite anode with platinum, the electrolysis pathway to functionalize indoles was changed (Scheme 16). Using compound **36** as the coupling partner, compound **37a** was obtained in 94% yield, whose structure was confirmed by X-ray diffraction. The *N*-protecting group of indole was crucial to the success of the dearomative coupling reaction. A tosyl-protecting group also facilitated the reaction to give the desired product in 91% yield (**37b**). Replacement of the electron-withdrawing *N*-protecting group of indole with methyl or the removal of this group shut down the reaction (**37c, 37d**). The bulky naphthyl group at the C3 position of indole could also form desired product **37e** with a crowded quaternary carbon in 83% yield, the structure of which was further confirmed by X-ray diffraction. This electrochemical transformation shows broad functional-group tolerance, accommodating Me, OMe, OCF_3, F, Cl, and I groups as well as heterocycles (**37f-37q**). Notably, the substrate scope can be successfully extended to 2-phenylbenzo[b]thiophene as a coupling partner, delivering the oxindole products in good yields (**37r**, 73%; **37s**, 78%).

They also performed detailed mechanistic investigations. Cyclic voltammetry experiments of benzofuran (**36a**) and indole (**34a**) with Pt or graphite as the working electrode showed that indole (**34a**) and benzofuran (**36a**) possess similar oxidation potentials with Pt as the anode. When using a graphite anode, however, **36a** displayed a much lower oxidation potential, while the value for **34a** did not shift appreciably. These results suggested that **36a** can be oxidized at the same time as or before **34a** simply by changing the anode materials under the same standard conditions, which provides the chance to develop different reactions. When adding $P(OEt)_3$ to the coupling reaction between **34a** and **36a**, compounds **38** and **39** were detected by HR-MS (Scheme 17a) [72, 73]. In addition, a trace mount of compound **40** was also isolated during the preparation process of compound **37a** (Scheme 17b) [53]. These results indicated that indole and benzofuran radical cation may be involved in this electrochemical transformation. In addition, when $H^{18}O$ was employed as a possible nucleophile, ^{18}O-labeled **37a'** was not detected by

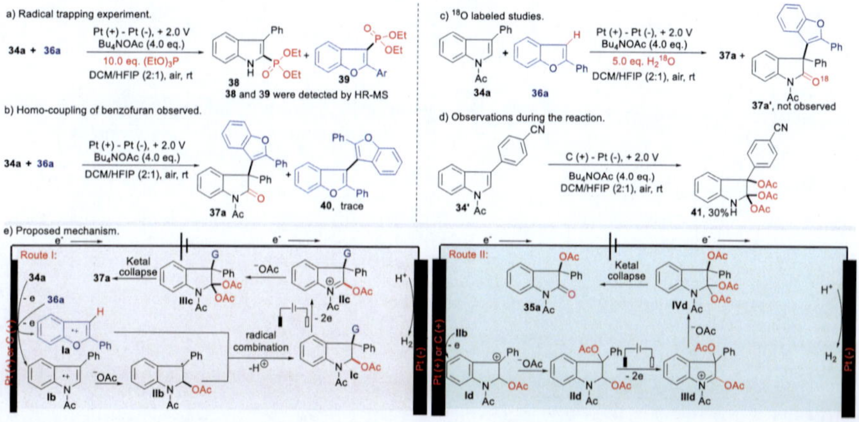

Scheme 17 Detailed mechanistic studies of the reaction in Scheme 16

HR-MS, ruling out residual H_2O in the solvent as the source of carbonyl in desired product (Scheme 17c). With **34′** as reactant, a triacetoxy product **41** was isolated in 30% yield, suggesting that acetate may be introduced in a stepwise manner and that the carbonyl group in desired product may be formed via hydrolysis of the ketal unit. Based on these findings, a plausible mechanism was proposed (Scheme 17d). When Pt was used as working electrode, compounds **34a** and **36a** were simultaneously oxidized to give radical cations **Ia** and **Ib** when using Pt as the working electrode. Intermediate **Ib** reacts with acetate anion via nucleophilic attack at the C2 position to deliver benzylic radical **IIb**, which undergoes a radical-radical cross-coupling with **Ia** to obtain intermediate **Ic**. **Ic** was further oxidized at the anode to give iminium cation **IIc**, which readily reacts with another acetate anion to form the unstable ketal **IIIc**. Product **37a** could be obtained via ketal collapse of **IIIc**. Benzofuran **36a** is oxidized first when using a graphite anode, however, forming an inseparable impurity. Subsequent formation and oxidation of **IIb** at the graphite anode forms benzylic cation **Id**. Similar anodic oxidation, nucleophilic attack by acetate anion, and ketal decomposition would then occur as above to form the final product **35a**.

Indeed, electrochemistry showed a unique ability to regulate the reactivity and chemoselectivity via tuning the applied potential, electrode materials, solvent, etc., enabling numerous elegant new methodologies [74]. Feng and coworkers found that indole and other heterocycles can undergo intermolecular dehydrogenative homocoupling and cross-coupling under electrochemical condition and that the coupling products could be further varied simply by extending the electrolysis time [75]. When they strictly limited the electrolysis time to 2 h, various indole derivatives, even without a *N*-protecting group, could undergo efficient homocoupling (Scheme 18). The reaction tolerated several functional groups, such as Me, F, Cl, Br, OCF_3, CN, CO_2Me, OCF_3, etc., at various positions on the benzene ring of the indole (**44a-44n**). In addition, various *N*-protected indoles, such as *N*-Me (**44o**), *N*-CO_2Et (**44p**), *N*-Ac (**44q**), and *N*-Ts (**44r**) units, also exhibited good efficiency for this electrochemical homocoupling, with acetyl providing the highest

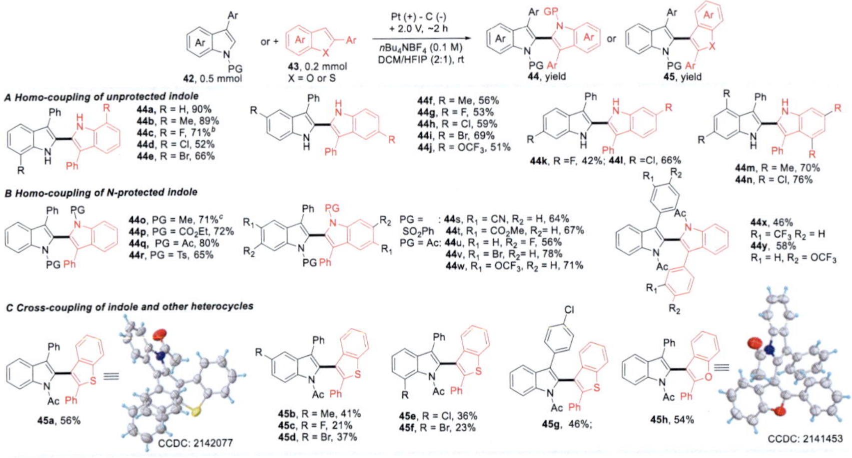

Scheme 18 Selected examples of electrochemical indole coupling reactions

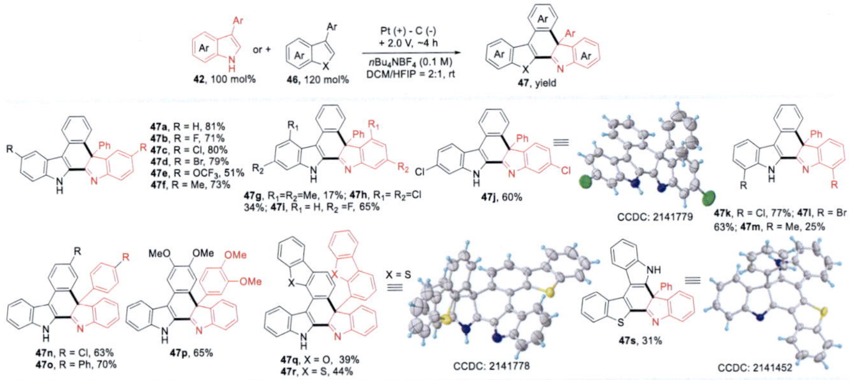

Scheme 19 Selected examples of electrochemical indole coupling cyclization reactions

efficiency (**44r**, 80%). Other *N*-Ts or *N*-Ac-protected indoles with electron-withdrawing substituents on the benzene ring gave products in good to acceptable yields (**44s-44y**). Notably, after screening different *N*-protected indoles, they found *N*-Ac-3-phenyl indole with different functionalities (**45a-45g**) could selectively cross-couple with 2-phenylbenzothiophene in 21–56% yield. Moreover, the synthetic utility of this electrochemical cross-coupling with indole was successfully expanded to benzofuran to give **45h** in 54% yield. The structure of **45a** and **45h** was further confirmed by X-ray crystallography.

By simply extending the reaction time to 4 h, Feng and coworkers achieved tandem C-C coupling and cyclization reaction to construct polycyclic scaffolds that possess greater structural diversity and complexity for a wide range of applications but are challenging to synthesize. As shown in Scheme 19, free indoles with various

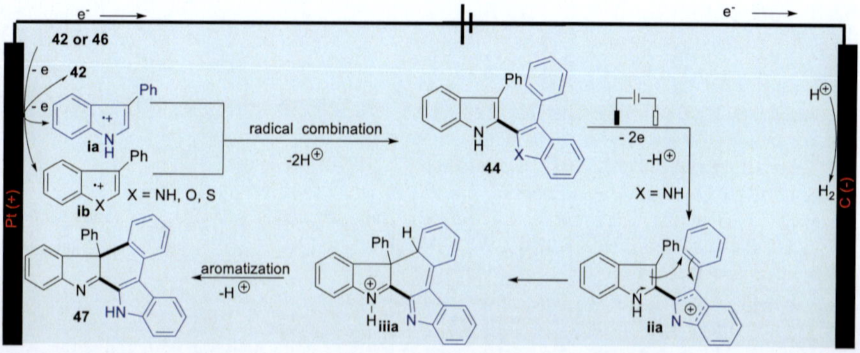

Scheme 20 The proposed reaction mechanism for time-dependent indole derivatization under electrochemical condition

substituents on the benzene ring afforded polycyclic products in good yields (**47a**-**47m**). The low yields of **47g** (17%), **47h** (34%), and **47m** (25%) were attributed solubility issues, with most of them precipitating during the transformation. Additionally, substrates bearing various arenes at C3 position were transformed into the desired products in good yields (**47a**-**47p**). Notably, polyaromatic products **47q** and **47r** with a crowded quaternary carbon center were successfully delivered in 39% and 44% yields, respectively. A cascade cross-coupling and cyclization between indole and benzothiophene units was also achieved to give product **47s** in 31% yield. The structures of **47j**, **47r**, and **47s** were further confirmed by X-ray crystallography.

In light of the outstanding versatility of these time-dependent electrochemical transformations, the authors performed several experiments to understand its mode of action. They first conducted cyclic voltammetry experiments for indole derivatives, benzofuran, benzothiophene, and homocoupling intermediates with a Pt working electrode. *N*-Ac-protected indole showed similar oxidation onset as benzofuran and benzothiophene, which indicated that simultaneous oxidation of these compounds on the electrode surface is possible. The homocoupling intermediate showed a higher oxidation potential that indole derivatives. Thus, after complete consumption of indole by electrolysis, the homocoupling intermediate is then oxidized. The proposed reaction mechanism is summarized in Scheme 20. Indoles **42** and other heterocycles **46** may be simultaneously oxidized on the surface of Pt anode to give aromatic radical cations **ia** and **ib**, which undergo homocoupling or cross-coupling to deliver radical-radical coupling product **44**. The indole dimer could be further oxidized to form the intermediate cation **iia**, which could initiate an intramolecular quasi-Michael addition with another indole unit as the nucleophile, generating cation **iiia**. The final cyclization product **47** would be obtained after aromatization. Meanwhile, cathodic reduction of protons forms H_2, obviating the need for sacrificial external chemical oxidants.

References

1. Katritzky AR (2004) Chem Rev 104:2125–2126
2. Fattorusso E, Taglialatela-Scafati O (2007) Modern alkaloids: structure, isolation, synthesis, and biology. Wiley
3. Taber DF, Tirunahari PK (2011) Tetrahedron 67:7195–7210
4. Shiri M (2012) Chem Rev 112:3508
5. Eicher T, Hauptmann S, Speicher A (2012) The chemistry of heterocycles: structure, reactions, syntheses, and applications.3rd edn. Wiley-VCH
6. Van Order RB, Lindwall HG (1942) Chem Rev 30:69–96
7. Humphrey GR, Kuethe JT (2006) Chem Rev 106:2875
8. Robinson B (1969) Chem Rev 69:227–250
9. Gassman PG, Van Bergen TJ, Gruetzmacher G (1973) J Am Chem Soc 95:6508–6509
10. Moody CJ, Swann E (1998) Synlett 1998:135–136
11. Carrera Jr GM, Sheppard GS (1994) Synlett:93–94
12. Brenner M, Mayer G, Terpin A, Steglich W (1997) Chem A Eur J 3:70–74
13. Larock RC, Yum EK (1991) J Am Chem Soc 113:6689–6690
14. Wender PA, White AW (1983) Tetrahedron 39:3767–3776
15. Wertjes WC, Southgate EH, Sarlah D (2018) Chem Soc Rev 47:7996–8017
16. Crich D, Banerjee A (2007) Acc Chem Res 40:151–161
17. Hang D, Song H, Qin Y (2011) Acc Chem Res 44:447–457
18. Bartoli G, Bencivenni G, Dalpozzo R (2010) Chem Soc Rev 39:4449–4465
19. Jung ME, Slowinski F (2001) Tetrahedron Lett 42:6835–6838
20. Zhu M, Zheng C, Zhang X, You S-L (2019) J Am Chem Soc 141:2636–2644
21. Mihelcic J, Moeller KD (2000) Tetrahedron 56:9527–9554
22. Yoshida J, Kataoka K, Horcajada R, Nagaki A (2008) Chem Rev 108:2265–2299
23. Kirste A, Schnakenburg G, Stecker F, Fischer A, Waldvogel SR (2010) Angew Chem Int Ed 49: 971–975
24. Morofuji T, Shimizu A, Yoshida J (2013) J Am Chem Soc 135:5000–5003
25. Francke R, Little RD (2014) Chem Soc Rev 43:2492–2521
26. Yan M, Kawamata Y, Baran PS (2017) Chem Rev 117:13230–13319
27. Sawamura T, Takahashi K, Inagi S, Fuchigami T (2012) Angew Chem Int Ed 51:4413–4416
28. Xiong P, Xu H-H, Xu H-C (2017) J Am Chem Soc 139:2956–2959
29. Horn EJ, Rosen BR, Baran PS (2016) ACS Cent Sci 2:302–308
30. Francke R (2014) Beilstein J Org Chem 10:2858–2873
31. Jiang Y, Xu K, Zeng C (2018) Chem Rev 118:4485–4540
32. Xiong P, Xu H-C (2019) Acc Chem Res 52:3339–3350
33. Hou Z-W, Mao Z-Y, Zhao H-B, Melcamu YY, Lu X, Song J, Xu H-C (2016) Angew Chem Ent Ed 55:9168–9173
34. Du P, Brosmer JL, Peters DG (2011) Org Lett 13:4072–4075
35. Tang S, Gao X, Lei A (2017) Chem Commun 53:3354–3356
36. Xu F, Li Y-J, Huang C, Xu H-C (2018) ACS Catal 8:3820–3824
37. Miller CP, Collini MD, Tran BD, Harris HA, Kharode YP, Marzolf JT, Moran RA, Henderson RA, Bender RHW, Unwalla RJ, Greenberger LM, Yardley JP, Abou-Gharbia MA, Lyttle CR, Komm BS (2001) J Med Chem 44:1654–1657
38. Cacchi S, Fabrizi G (2005) Chem Rev 105:2873–2920
39. Yuan G, Zhao H, Liu X, Hasanali ZS, Zou Y, Levine A, Wang D (2009) Angew Chem Int Ed 48:9680–9644
40. Zhou N, Zhao J, Sun C, Lai Y, Ruan Z, Feng P (2021) J Org Chem 86:16059–16067
41. Rao SN, Babu KS (2011) Org Commun 4:105–111
42. Comber MF, Moody CJ (1992) Synthesis 1992:731
43. Harrison JR, Moody CJ (2003) Tetrahedron Lett 44:5189–5191
44. Yamada K, Yamada F, Shiraishi T, Tomioka S, Somei M (2009) Heterocycles 77:635–645

45. Poirier M, Goudreau S, Poulin J, Savoie J, Beaulieu PL (2010) Org Lett 12:2334–2337
46. Beukeaw D, Udomsasporn K, Yotphan S (2015) J Org Chem 80:3447–3454
47. Shrestha A, Lee M, Dunn AL, Sanford MS (2018) Org Lett 20:204–207
48. Yang Q-L, Wang X-Y, Lu J-Y, Zhang L-P, Fang P, Mei T-S (2018) J Am Chem Soc 140: 11487–11494
49. Wu Y-C, Jiang S-S, Song R-J, Li J-H (2019) Chem Commun 55:4371–4374
50. Peng X, Zhao J, Ma G, Wu Y, Hu S, Ruan Z, Feng P (2021) Green Chem 23:8853–8858
51. Ito A, Yokoyama Y, Aihara R, Fukui K, Eguchi S, Shizu K, Sato T, Tanaka K (2010) Angew Chem Int Ed 49:8205–8208
52. Kim M-H, Park Y, Jeong B-S, Park H-G, Jew S-S (2010) Org Lett 12:2826–2829
53. Bering L, Paulussen FM, Antonchick AP (2018) Org Lett 20:1978–1981
54. Roche Jr SP, Porco JA (2011) Angew Chem Int Ed 50:4068–4093
55. Zi W, Zuo Z, Ma D (2015) Acc Chem Res 48:702–711
56. Roche SP, Youte Tendoung J-J, Tréguier B (2015) Tetrahedron 71:3549–3591
57. Abou-Hamdan H, Kouklovsky C, Vincent G (2020) Synlett 31:1775–1788
58. Shen SJ, Zhu CL, Lu DF, Xu H (2018) ACS Catal 8:4473–4482
59. Zhang Z, Li X, Song M, Wan Y, Zheng D, Zhang G, Chen G (2019) J Org Chem 84:12792–12799
60. Duan K, Shi H-J, Wang L-X, Li S-S, Xu L-B, Xiao J (2020) Org Chem Front 7:2511–2517
61. Jayaraman A, Castro LCM, Desrosiers V, Fontaine FG (2018) Chem Sci 9:5057–5063
62. Lautens M, Pultar FA (2016) Synfacts 12:1044–1044
63. Zhao J-C, Yu S-M, Liu Y, Yao Z-J (2013) Org Lett 15:4300–4303
64. Bandini M (2013) Org Biomol Chem 11:5206–5212
65. Schotten C, Nicholls TP, Bourne RA, Kapur N, Nguyen BN, Willans CE (2020) Green Chem 22:3358–3375
66. Liu K, Tang S, Huang P, Lei A (2017) Nat Commun 8:775
67. Jeffrey CS, Barnes KL, Eickhoff JA, Carson CR (2011) J Am Chem Soc 133:7688–7691
68. Fishman JM, Kiessling LL (2013) Angew Chem Int Ed 52:5061–5064
69. Li L, Luo S (2018) Org Lett 20:1324–1327
70. Wu J, Dou Y, Guillot R, Kouklovsky C, Vincent G (2019) J Am Chem Soc 141:2832–2837
71. Peng X, Wen L, Ning Z, Zhang Z, Sun C, Tang Y, Feng P (2022) Org Chem Front 9:3800–3806
72. Yurko EO, Gryaznova TV, Kholin KV, Khrizanforova VV, Budnikova YH (2018) Dalton Trans 47:190–196
73. Long H, Huang C, Zheng Y-T, Li Z-Y, Jie L-H, Song J, Zhu S, Xu H-C (2021) Nat Commun 12:6629
74. Heard DM, Lennox AJJ (2020) Angew Chem Int Ed 59:18866–18884
75. Ning Z, Zhang Z, Yan Q, Zhou N, Wen L, Peng X, Tang Y, Feng P (2022) Sci China Chem 65: 1962–1967

Top Heterocycl Chem (2024) 61: 23–36
https://doi.org/10.1007/7081_2024_76

Published online: 24 August 2024

Construction and Functionalization of Heterocylces via Electrochemical C–H Functionalization

Ping Fang, Jian-Feng Guo, and Tian-Sheng Mei

Contents

P. Fang, J.-F. Guo, and T.-S. Mei (✉)
State Key Laboratory of Organometallic Chemistry, Shanghai Institute of Organic Chemistry, University of Chinese Academy of Sciences, Chinese Academy of Sciences, Shanghai, P. R. China
e-mail: mei7900@sioc.ac.cn

Abstract This chapter mainly introduces our research in constructing heterocycles (α-pyrones, α-pyridones, and dihydroquinazolinones) through the functionalization of C–H bonds and in derivatizations (amination, halogenation, alkylation, alkynylation, alkenylation, cyanation, phosphonylation, and annulation) of existing heterocycles through the functionalization of C–H bonds. These advancements can provide a solid foundation for future progress in this field.

Keywords Electrochemical C–H Functionalization · Heterocylces

1 General

Heterocyclic compounds are a class of molecules with rich structures and properties, which have wide applications and developments [1]. In biochemistry, many natural products such as vitamins and growth hormones are composed of heterocycles, which have high biological activity and therapeutic effects and are widely used in the pharmaceutical field [2]. Heterocyclic compounds also have diverse physical and chemical properties [3–5], making them important raw materials for synthesizing new chemicals. For example, various heterocyclic polymers have potential applications owing to their unique interactions with light, electricity, and magnetism. In addition, heterocyclic compounds are also widely used in the field of environmental protection. Many toxic and harmful organic substances [6, 7], such as benzopyrene and polycyclic aromatic hydrocarbons, are often degraded and decomposed through the cyclization, oxidation, and reduction reactions of heterocyclic compounds, thereby improving the environment.

The application and development of heterocyclic compounds are not limited to a single field, but their wide and diverse properties and applications provide chemists and scientists with infinite exploration and development space, improving our quality of life.

However, due to the complex structure of heterocyclic compounds, multi-step reactions are often required for synthesis, which poses significant difficulties and challenges. Constructing or derivatizing heterocycles by directly functionalizing the C–H bonds of organic small molecules is therefore one of the most efficient strategies. By leveraging electrochemistry, not only can stoichiometric oxidants be avoided in the oxidation of C–H bonds, but this platform can also be used to regulate the selectivity of the reaction. Therefore, electrochemically promoted synthesis and derivatization via C–H functionalizations are highly efficient new approaches to afford various heterocyclic compounds.

2 Heterocycle Synthesis via Electrochemical C–H Functionalization

2.1 Construction of α-Pyrones via Electrooxidative Iridium-Catalyzed Regioselective Annulation

The Mei Group developed an electrochemically driven, Cp*Ir(III)-catalyzed regioselective annulation to construct α-pyrones (Scheme 1) [8, 9]. Acrylic acids and benzoic acids were suitable for the reaction with various internal alkynes. Both **M1** and **M2** were isolated and could be used as catalysts for the reaction. Treatment of **M2** with representative chemical oxidants, such as Ag_2CO_3 or $Cu(OAc)_2$, did not give significant amounts of product. Mechanistic studies demonstrated that anodic oxidation was crucial to turn over the diene-Ir(I) complex to an Ir(III) species with concomitant release of the product. The regioselectivity when using unsymmetrical alkynes was good to excellent depending on the relative steric hindrance, electronic properties, or binding affinity with the iridium catalyst of the two substituents. The steric bias of the system places the smaller alkyne substituent near the β-carbon of the acrylic or benzoic acid in **M1** and thus forms the new C–C bond to the alkynyl carbon attached to this group. Reversed regioselectivity was observed for aryl, trifluoromethyl alkynes, however, with the nucleophilic Ir–C(sp^2) inserting into the more electrophilic alkynyl carbon (distal to the smaller CF_3 group). Lastly, tertiary propargylic alcohols deliver products with the alcohol group proximal to the new C–O bond, likely owing to coordination of the alcohol to the Ir center before C–C bond formation.

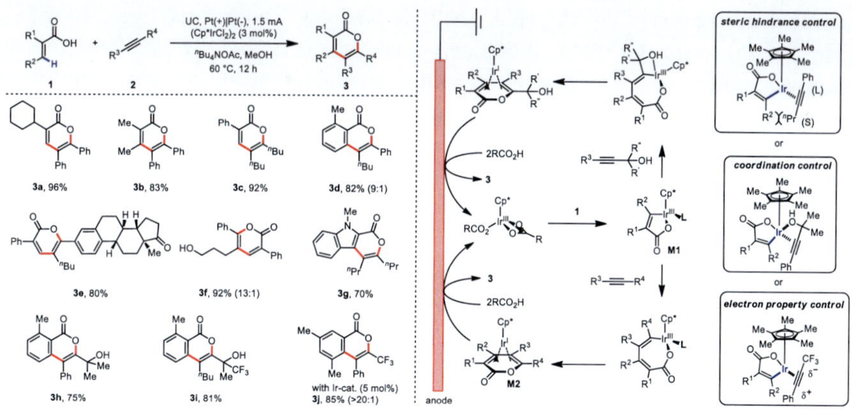

Scheme 1 The construction of α-pyrones via electrooxidative Ir-catalyzed annulation

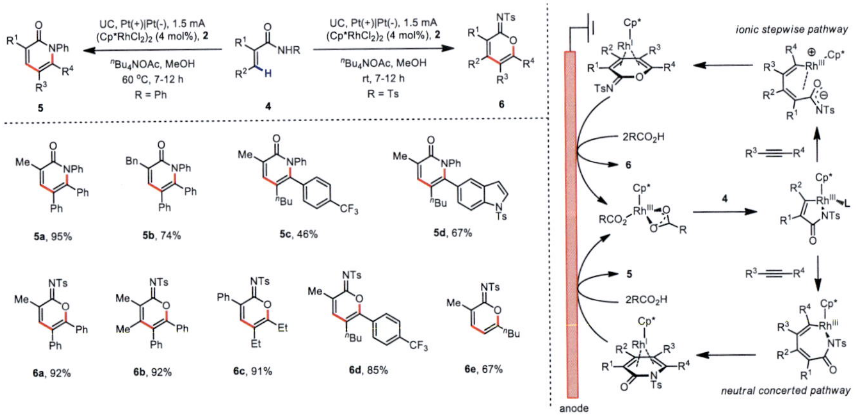

Scheme 2 The construction of α-pyridones and α-pyrones via electrooxidative Rh-catalyzed annulation

2.2 *Construction of α-Pyridones and α-Pyrones via Rh-Catalyzed Electrochemical Vinylic C–H Annulation of Acrylamides with Alkynes*

Under rhodium catalysis, divergent constructions of α-pyridones and cyclic imidates were developed by the Mei Group (Scheme 2) [10]. Using *N*-phenyl acrylamides as substrates, the protocol afforded the α-pyridones. With *N*-tosyl acrylamides as substrates, on the other hand, it afforded the cyclic imidates. DFT calculations elucidated the reaction mechanism and origins of substituent-controlled chemoselectivity. It might undergo either a classical neutral, concerted reductive elimination to produce α-pyridones, or an ionic, stepwise pathway to produce cyclic imidates. The neutral, concerted reductive elimination has barrier of 31.6 kcal/mol with *N*-tosyl acrylamide as substrate, which is unfeasible under the experimental conditions. For *N*-phenyl acrylamide, it could undergo the classic neutral concerted reductive elimination pathway with an accessible barrier of 21.9 kcal/mol, whereas the alternative ionic stepwise pathway was significantly less favorable due to the unstable zwitterionic species.

2.3 *Construction of Dihydroquinazolinones via Electrochemical Rhodium-Catalyzed C–H Functionalization*

Assisted by a weakly coordinating amide, an electrochemical rhodium-catalyzed C–H amination and an aminative cyclization of arenes were developed (Scheme 3) [11]. By controlling the stoichiometry of electric charge applied, the system could

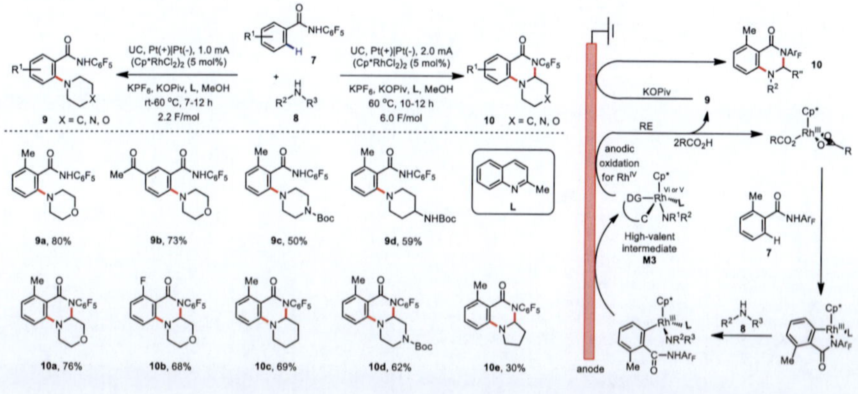

Scheme 3 Divergent synthesis of aryl amines and dihydroquinazolinones via electrochemical rhodium-catalyzed C–H functionalization

afford 2-aminobenzamides or dihydroquinazolinones selectively. The latter variant provides an especially simple and robust protocol to construct dihydroquinazolinones. Mechanistic studies demonstrated that an electrooxidatively induced reductive elimination of high-valent rhodium intermediate **M3** was probably involved in the formation of aryl amine **9**, which is generated in both protocols. Dihydroquinazolinones are produced by continued electrolysis because product **9** is oxidized further under electrochemical conditions, with KOPiv assisting as a base, to afford the cyclization product **10**.

3 C–H Functionalization Toward Heterocycles

3.1 Copper-Catalyzed Electrochemical C–H Amination

Copper-catalyzed electrochemical C–H aminations of (hetero)arenes were reported for the first time in 2018 (Scheme 4) [12]. Various cycloamines were suitable for the reaction, such as morpholine, thiomorpholine, piperidine, and mono-*N*-protected piperazine. This amination protocol could be applied not only to aryl C–H bonds but also to heteroaryl C–H bonds, with substrates such as pyridine and quinolone proving successful. A picolinamide directing group proved optimal. Using a quaternary ammonium iodide salt as a redox mediator was indispensable for this transformation. Mechanistic studies demonstrated that the reaction proceeds via a single-electron-transfer (SET) process and that a high-valent Cu(III) species is likely involved.

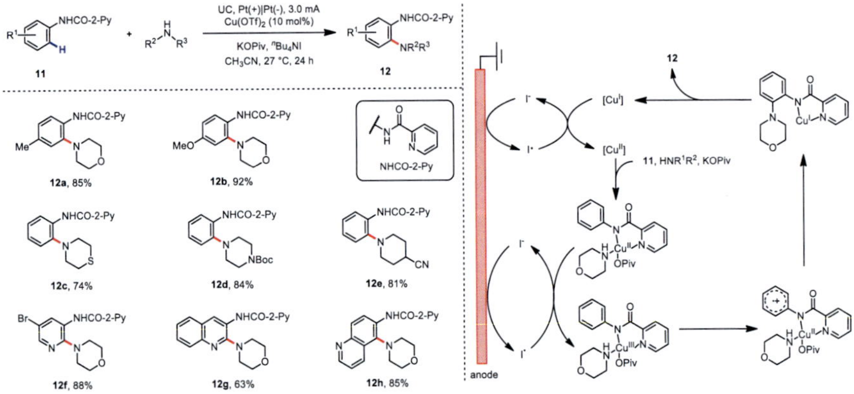

Scheme 4 Copper-catalyzed electrochemical C–H amination

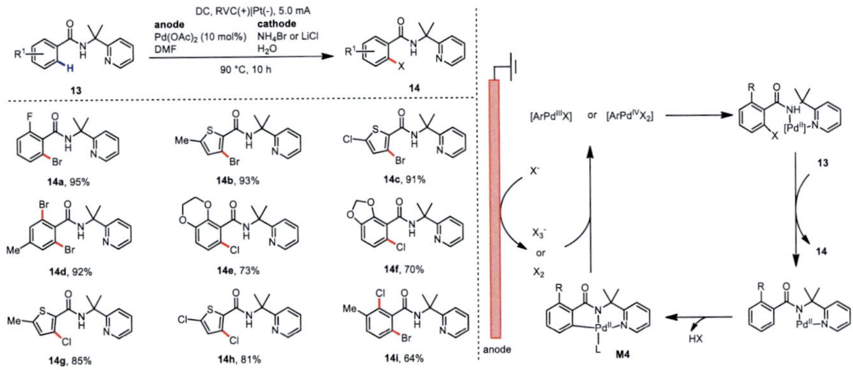

Scheme 5 Palladium-catalyzed electrochemical C–H halogenation

3.2 *Palladium-Catalyzed Electrochemical C–H Halogenation*

Palladium-catalyzed electrochemical C–H halogenations of benzamide derivatives in divided cells were also developed (Scheme 5) [13, 14]. In these reactions, cheap and readily available NH_4Br or LiCl served as both halogenating reagent and electrolyte. This catalytic electrochemical bromination protocol is suitable for thiophene substrates as heterocycles. Additionally, the sequential bromination and chlorination of C(sp^2)–H bond constructs highly functionalized aromatic carboxylic acid derivatives. Palladacycle **M4** was synthesized, and it was an active catalyst for the halogenation reaction, which indicated that the reactive palladacyle for the reaction could be monomeric.

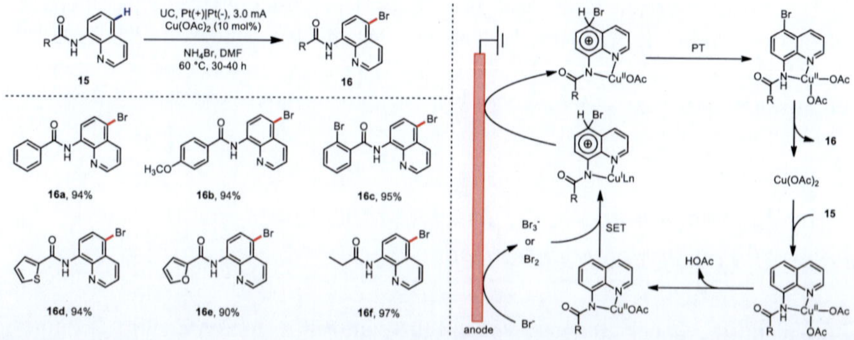

Scheme 6 Copper-catalyzed electrochemical C–H selective bromination of 8-aminoquinoline amide

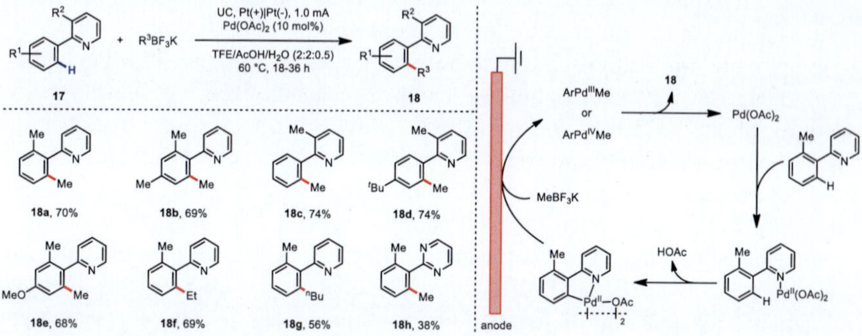

Scheme 7 Palladium-catalyzed electrochemical C–H alkylation

3.3 *Copper-Catalyzed Electrochemical C–H Selective Bromination*

An environmentally friendly protocol for copper-catalyzed bromination of quinoline at the C5 site by anodic oxidation was developed (Scheme 6) [15], affording remote C–H bromination products. Low-cost NH_4Br served as both brominating reagent and electrolyte. Cyclic voltammetry (CV) experiments were consistent with the proposed catalytic cycle.

3.4 *Palladium-Catalyzed Electrochemical C–H Alkylation*

C–H bonds of arenes with a directing group (pyridine or pyrimidine) could be electrochemically alkylated under palladium catalysis (Scheme 7) [16]. Potassium trifluoroalkylborates served as alkylating reagents, using a combination TFE/AcOH/

H_2O as solvent without additional electrolyte. This protocol represents an environmentally friendly electrochemical synthesis. Mechanistic studies demonstrated that the alkylation involves a high-valent Pd(III) or Pd(IV) center prior to reductive elimination.

3.5 *Electrochemical C–H Functionalization of Tetrahydroisoquinolines*

Via merging copper catalysis and electrochemistry, an asymmetric Shono-type oxidative cross-coupling was developed by the Mei group (Scheme 8) [17]. It could afford C1-alkynylated tetrahydroisoquinolines with good to excellent enantioselectivity. The use of TEMPO as a co-catalytic redox mediator is crucial not only for oxidizing a tetrahydroisoquinoline to an iminium ion (**M5**), but also for decreasing the oxidation potential of the anode.

Employing 4-dimethylaminopyridine as an organocatalyst and TEMPO/NaBr as catalytic mediators, alkenylation of tetrahydroisoquinolines was developed via merging Shono oxidation and the Morita–Baylis–Hillman reaction (Scheme 9) [18]. The use of TEMPO and NaBr as mediators was crucial to avoid the decomposition of products by lowering the oxidation potential of the anode. On the basis of these mechanistic experiments and literature precedent, there might be two ways to simultaneously produce key intermediates **M5** (path A: $TEMPO^+$ as mediator, and path B: Br_3^- as mediator. Path A was identified as the major path). Then, the asymmetric alkenylation of tetrahydroisoquinolines was investigated. Employing quinine as a chiral catalyst, it could afford 2-(2-phenyl-1,2,3,4-tetrahydroisoquinolin-1-yl)acrylaldehyde **23 g** with 90% yield and 78% ee (Scheme 10) [19].

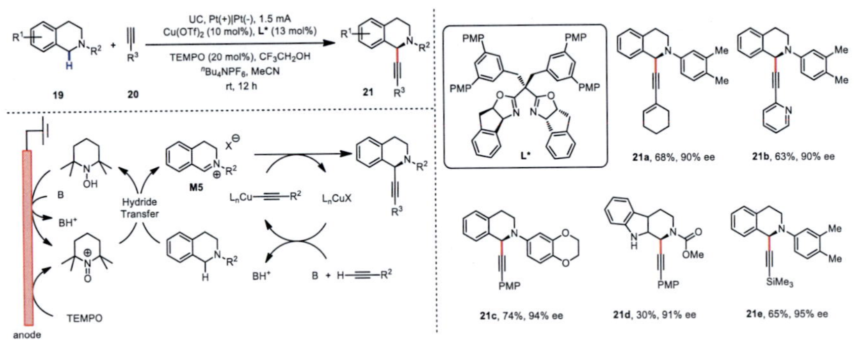

Scheme 8 Copper-catalyzed electrochemical C–H alkynylation

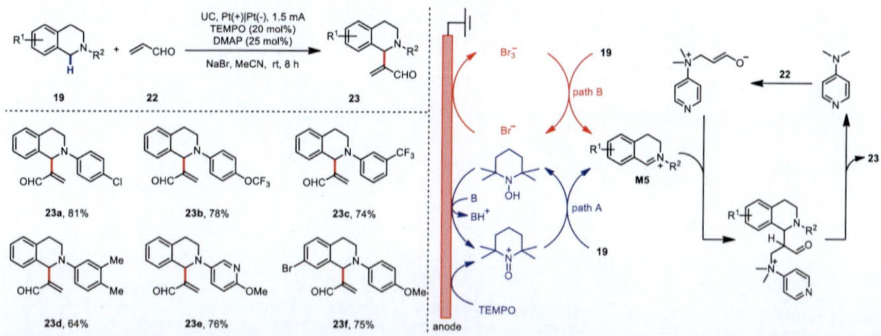

Scheme 9 TEMPO-catalyzed electrochemical C–H alkenylation

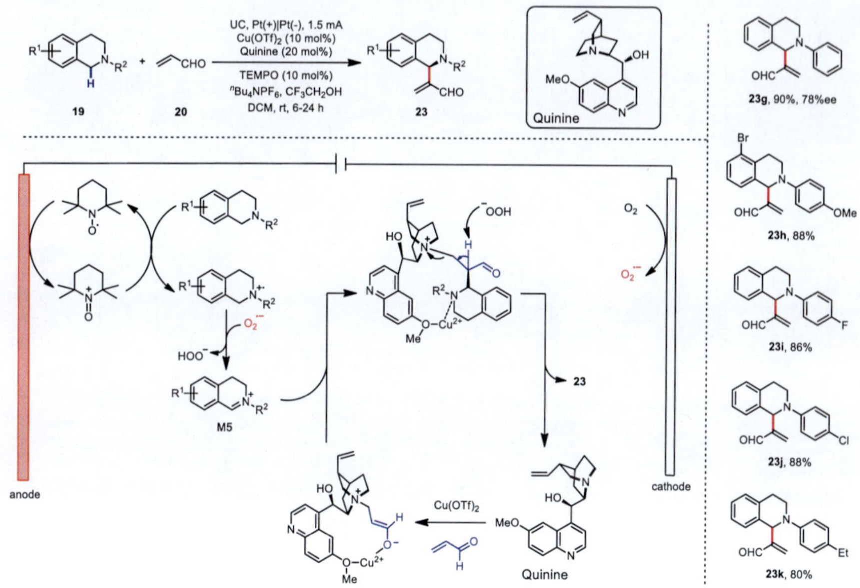

Scheme 10 Asymmetric electrochemical C–H alkenylation

3.6 *TEMPO-Catalyzed Electrochemical C–H Functionalization of Cyclic Amines*

Metal-free electrochemical oxidative cyanation and phosphonylation reactions were also developed (Scheme 11) [20]. TEMPO lowered the anodic potential needed to engage the substrate and avoided over-oxidation of electron-rich aromatic amines under electrochemical conditions.

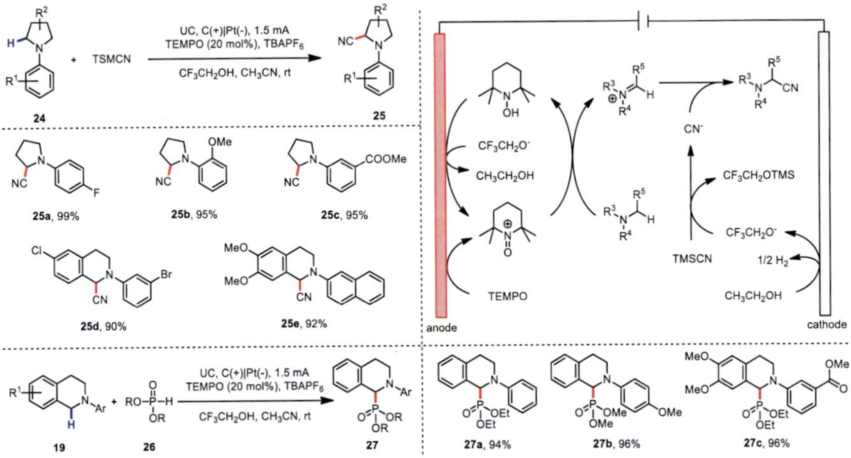

Scheme 11 TEMPO-catalyzed electrochemical C–H functionalization of cyclic amines

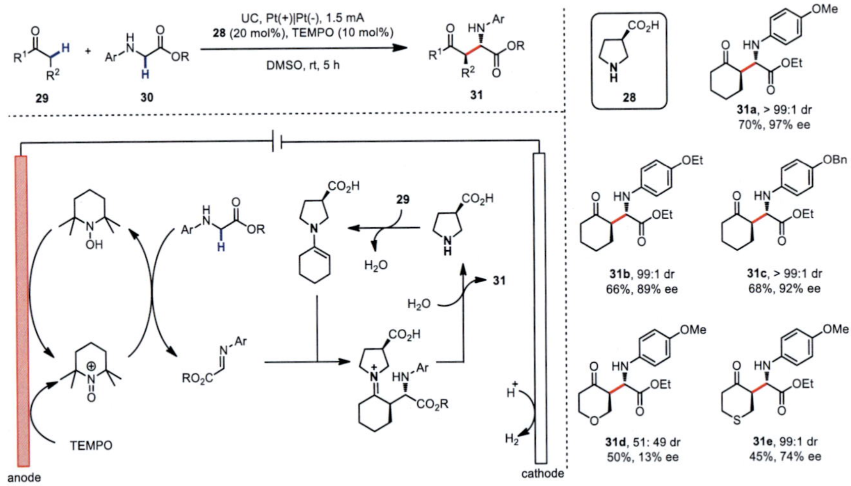

Scheme 12 Electrooxidative asymmetric coupling of secondary acyclic amines with ketones

3.7 *Electrooxidative Asymmetric C–C Coupling of Secondary Acyclic Amines with Ketones*

An electrochemical asymmetric C–C coupling of secondary acyclic amines (*N*-arylglycinate esters) with ketones via a Shono-type oxidation was developed (Scheme 12) [21]. The protocol was also suitable for cyclic ketones containing heteroatoms. The addition of TEMPO as a redox mediator could selectively oxidize the substrate rather than the product, although their oxidation potential difference is subtle. Mechanistic studies suggest that proton-mediated racemization of the product is prevented by the reduction of protons at the cathode.

3.8 *Electrochemical Rhodium-Catalyzed Enantioselective C–H Annulation with Alkynes*

An electrochemically tuned, rhodium-catalyzed enantioselective C–H annulation of (alkylidene)pyrazolones with alkynes was developed (Scheme 13) [22], which afforded various spiropyrazolones in good yields and enantioselectivities in an undivided cell at room temperature. Notably, unsymmetrical alkylarylacetylenes are also efficiently converted to their corresponding spiroannulation products with excellent regioselectivity and good enantioselectivity.

4 Summary

Two electrochemically promoted approaches to heterocyclic synthesis and derivatization via C–H bond functionalizations were developed by the Mei Group. α-Pyrones, α-pyridones, and dihydroquinazolinones could be constructed via transition-metal-catalyzed functionalization of C–H bonds. Furthermore, various electrochemically enabled functionalizations of heterocycle C–H bonds were realized, such as amination, halogenation, alkylation, alkynylation, alkenylation, cyanation, phosphonylation, and spiroannulation (though these transformations also perform C–H functionalization on substrates beyond heterocycles). These

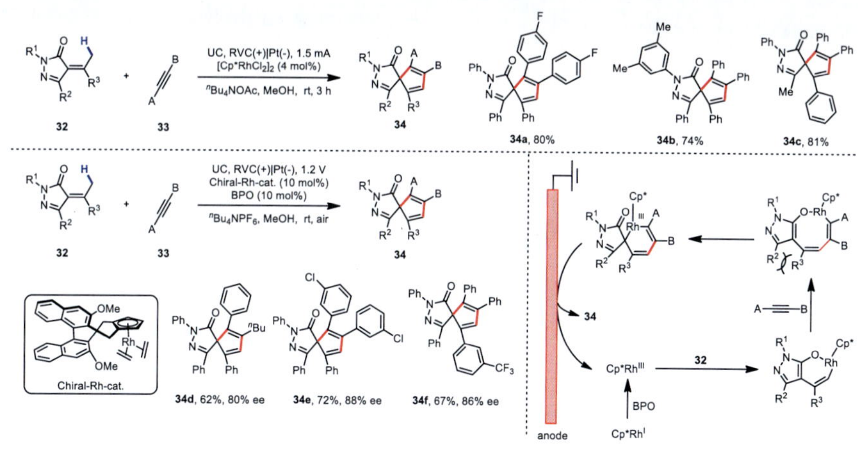

Scheme 13 Rh-catalyzed electrochemical C–H annulation

research advancements thus showcase the potential of electrochemistry for both the construction and functionalization of heterocylces.

References

1. Katritzky AR, Rees CW (2010) Comprehensive heterocyclic chemistry III. Elsevier Science
2. Kabir E, Uzzaman M (2022) A review on biological and medicinal impact of heterocyclic compounds. Results Chem 4:100606
3. Qian ZG, Pang ZZ, Li ZX, He MH, Liu JG, Fan L, Yang SY (2002) Photoimageable polyimides derived from α,α-(4-amino-3,5-dimethylphenyl)phenylmethane and aromatic dianhydride. J Polym Sci A Polym Chem 40:3012–3020
4. Feng L, Xie Y (2015) Heterocyclic polymers: synthesis, properties, and applications. CRC Press
5. Fuchigami T, Atobe M (2018) Heterocyclic polymers. Springer
6. Yang Y, Chen B, Shen S (2015) Heterocyclic compounds as a potent tool for the sustainability of the environment: a review. Chem Eng Technol 38(6):901–910
7. Gupta R, Sharma D (2019) Heterocyclic compounds: a green tool for environmental protection. Curr Res Green Sustain Chem 2:100011
8. Yang QL, Xing YK, Wang XY, Ma HX, Weng XJ, Yang X, Guo HM, Mei TS (2019) Electrochemistry-enabled Ir-catalyzed Vinylic C−H functionalization. J Am Chem Soc 141(48):18970–18976
9. Yang QL, Jia HW, Liu Y, Xing YK, Ma RC, Wang MM, Qu GR, Mei TS, Guo HM (2021) Electrooxidative iridium-catalyzed regioselective annulation of benzoic acids with internal alkynes. Org Lett 23(4):1209–1215
10. Xing YK, Chen XR, Yang QL, Zhang SQ, Guo HM, Hong X, Mei TS (2021) Divergent rhodium-catalyzed electrochemical vinylic C–H annulation of acrylamides with alkynes. Nat Commun 12:930–938
11. Xing YK, Wang ZH, Fang P, Ma C, Mei TS (2023) Divergent synthesis of aryl amines and dihydroquinazolinones via electrochemistry-enabled rhodium-catalyzed C–H functionalization. Sci China Chem 66(10):2863–2870
12. Yang QL, Xing YK, Lu JY, Zhang LP, Fang P, Mei TS (2018) Copper-catalyzed electrochemical C−H amination of arenes with secondary amines. J Am Chem Soc 140(36):11487–11494
13. Yang QL, Wang XY, Wang TL, Yang X, Liu D, Tong XF, Wu XY, Mei TS (2019) Palladium-catalyzed electrochemical C−H bromination using NH_4Br as the brominating reagent. Org Lett 21(8):2645–2649
14. Yang QL, Wang XY, Weng XJ, Yang X, Xu XT, Tong XF, Fang P, Wu XY, Mei TS (2019) Palladium-catalyzed ortho-selective C—H chlorination of arenes using anodic oxidation. Acta Chim Sin 77(9):866–873
15. Yang X, Yang QL, Wang XY, Xu HH, Mei TS, Huang Y, Fang P (2019) Copper-catalyzed electrochemical selective bromination of 8-aminoquinoline amide using NH4Br as the brominating reagent. J Org Chem 85(5):3497–3507
16. Yang QL, Li CZ, Zhang LW, Li YY, Tong XF, Wu XY, Mei TS (2019) Palladium-catalyzed electrochemical C−H alkylation of arenes. Organometallics 38(6):1208–1212
17. Gao PS, Weng XJ, Wang ZH, Zheng C, Sun B, Chen ZH, You SL, Mei TS (2020) Cu^{II}/TEMPO-catalyzed enantioselective $C(sp^3)$–H alkynylation of tertiary cyclic amines through Shono-type oxidation. Angew Chem Int Ed 59(35):15254–15259
18. Liu HL, He Z, Wang NN, Fang P, Mei TS (2023) $C(sp^3)$−H Alkenylation of tetrahydroisoquinolines via merging electrochemistry and organocatalysis. Org Lett 25(4): 608–613

19. He Z, Liu HL, Wang ZH, Jiao KJ, Li ZM, Li ZJ, Fang P, Mei TS (2023) C(sp^3)–H aerobic alkenylation of tetrahydroisoquinolines via organic electrosynthesis. J Org Chem 88(9): 6203–6208
20. Gao JQ, Weng XJ, Ma C, Xu XT, Fang P, Mei TS (2021) Electrochemical 2,2,6,6-tetramethylpiperidinyl-N-oxyl (TEMPO)-mediated alpha-cyanation and phosphonylation of cyclic amines with metal-free conditions. Chin J Org Chem 41(8):3223–3234
21. Wang ZH, Gao PS, Wang X, Gao JQ, Xu XT, He Z, Ma C, Mei TS (2021) TEMPO-enabled electrochemical enantioselective oxidative coupling of secondary acyclic amines with ketones. J Am Chem Soc 143(38):15599–15605
22. Huang YQ, Wu ZJ, Zhu L, Gu Q, Lu X, You SL, Mei TS (2022) Electrochemical rhodium-catalyzed enantioselective C–H annulation with alkynes. CCS Chem 4(9):3181–3189

Top Heterocycl Chem (2024) 61: 37–54
https://doi.org/10.1007/7081_2024_78

Published online: 24 August 2024

Electrochemical Oxidative Cross-Coupling and Annulation for Functionalized Heterocycles

Xinwei Hu and Zhixiong Ruan

Contents

Abstract This chapter outlines the construction of high-value nitrogen-containing or selenium-substituted heterocyclic compounds under direct electrolysis conditions. It begins by introducing the fundamental principles of organic electrochemical synthesis and provides a detailed review of the research progress by Ruan's group in the electrochemical synthesis of nitrogen-containing or selenium-substituted heterocyclic compounds. Utilizing various inexpensive and readily available reaction substrates, a series of efficient and environmentally friendly intermolecular cyclization reactions have been developed using direct electrolysis methods, including dehydrogenative coupling reactions and cyclization strategies. Compared to traditional methods, this strategy offers mild conditions and eliminates the need for additional oxidants and transition metal catalysts. The novel cyclization reactions achieved through electrochemistry provide a more efficient and environmentally friendly synthetic approach, avoiding the generation of significant by-products and enabling the synthesis of heterocyclic compounds that are challenging to access via traditional methods from simple substrates without prefunctionalization. Moreover,

X. Hu and Z. Ruan (✉)
School of Pharmaceutical Sciences, Guangzhou Medical University, Guangzhou, P.R. China
e-mail: xinweihu@gzhmu.edu.cn; zruan@gzhmu.edu.cn

the later part of this chapter demonstrates the modification of these synthetic methods on peptide molecules to illustrate their practical applications.

Keywords Cross-coupling · Cyclization · Electrochemical oxidation · Heterocycles · Nitrogen · Selenium

1 Introduction

It is well-known that N-containing heterocycles are a highly important class of organic compounds, widely present in biologically active natural products, organic materials, agrochemicals, and pharmaceuticals. Consequently, the quest for more efficient and greener synthetic routes [1] to construct heterocyclic compounds through novel cyclization reactions has garnered increasing attention from organic chemists. In recent years, electrochemical organic synthesis has made remarkable progress in the construction of nitrogen-containing heterocyclic compounds [2]. This chapter particularly focuses on regioselective reactions achieved through direct electrolysis as a strategy. Emphasis is placed on three different pathways: oxidative dehydrogenative coupling, Se-substituted heterocyclic cyclization reactions, and other cyclization reactions, to promote the formation of heterocyclic compounds in a chemoselective manner.

2 Oxidative Dehydrogenative Coupling Reaction

2.1 *Csp^2–H Dehydrogenative Coupling*

After conducting a literature review, it was observed that the utilization of electrochemical synthesis strategies to form C–X (C, N, Se) bonds has been extensively explored. Among these, the most prevalent reactions involve oxidative dehydrogenative coupling, where electrons are lost at the anode surface to form radical intermediates, which then further couple to yield products. At the same time, protons are reduced to hydrogen on the cathode surface. Indoles and their derivatives are widely present in natural products and drug molecules due to their significant biological activity, and their involvement in coupling reactions can generate a plethora of indole scaffold derivatives. Leonori et al. [3] first achieved the photocatalytic use of organic dye Eosin Y as a photoredox catalyst to promote the formation of amidyl radicals and further coupling with (hetero)aromatics. Regarding substrates, complex compounds remain applicable, demonstrating high functional group tolerance (Scheme 1a). Furthermore, Ackermann et al. demonstrated the feasibility of the electrocatalytic cross-coupling reaction between indole aromatic

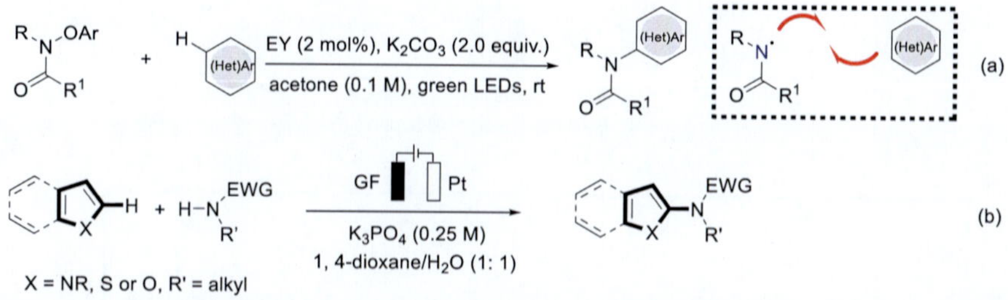

Scheme 1 Formation of amidyl radicals for (hetero)arene functionalization

heterocycles and sulfonamides [4]. Using electricity to release hydrogen gas as a by-product in an oxidative process, potassium phosphate as an electrolyte, and 1,4-dioxane/water (1:1) as a mixture solvent, the dehydrogenative coupling product was obtained after reacting for 20 h (Scheme 1b). The authors proposed the involvement of sulfonamidyl radicals, which further attacked indoles or heterocyclic compounds to achieve C − N bond coupling through dearomatization. The applicable aromatic heterocycles included not only indoles, but also benzothiophene and benzofuran. Finally, cyclic voltammetry and control experiments confirmed the presence of sulfonamidyl radicals.

Building upon previous work, Ruan et al. reported a method for selectively constructing C–N bonds by electrochemical oxidative cross-coupling of five-membered heteroaromatic compounds with aniline derivatives by C–H amination at the C2 or C3 positions of the heteroarene [5] (Scheme 2a). In contrast to the reaction reported by Ackermann's group, this system involves the reaction of arylamines with indole heterocycles, the former of which, compared to alkylamines, form more stable nitrogen radicals. This electrochemical synthetic method enables facile access to various synthetically useful heteroaromatic derivatives, tolerating multiple functional groups, and is applicable to gram-scale synthesis. Preliminary mechanistic studies indicated that a radical cross-coupling process between the *N*-centered radical and the indole radical cation is the key step in this amination reaction. Furthermore, Waldvogel et al. [6] utilized electrochemical conditions to generate stable nitrogen radicals by varying the reaction parameters and feedstock solubility modulation to enable the generation of N–N bonds via selective dehydrogenation (Scheme 2b). This reaction, which employs a sustainable strategy without transition metals or external oxidants, exhibits wide generality. Additionally, the authors evaluated the biological activity of benzimidazole derivatives, demonstrating antibacterial activity against human cancer cells and cytotoxicity. Some compounds also exhibited excellent photochemical properties and were successfully applied in subcellular imaging.

In the following years, the electrocatalytic dehydrogenative coupling reactions of indole derivatives with *N*-radicals continued to be investigated [7] (Scheme 3a). Selecting 3-phenylindole and 4-bromopyrazole as model substrates, with the cell voltage controlled at 2.5 V, a complete indolation at the N1 position of pyrazole with

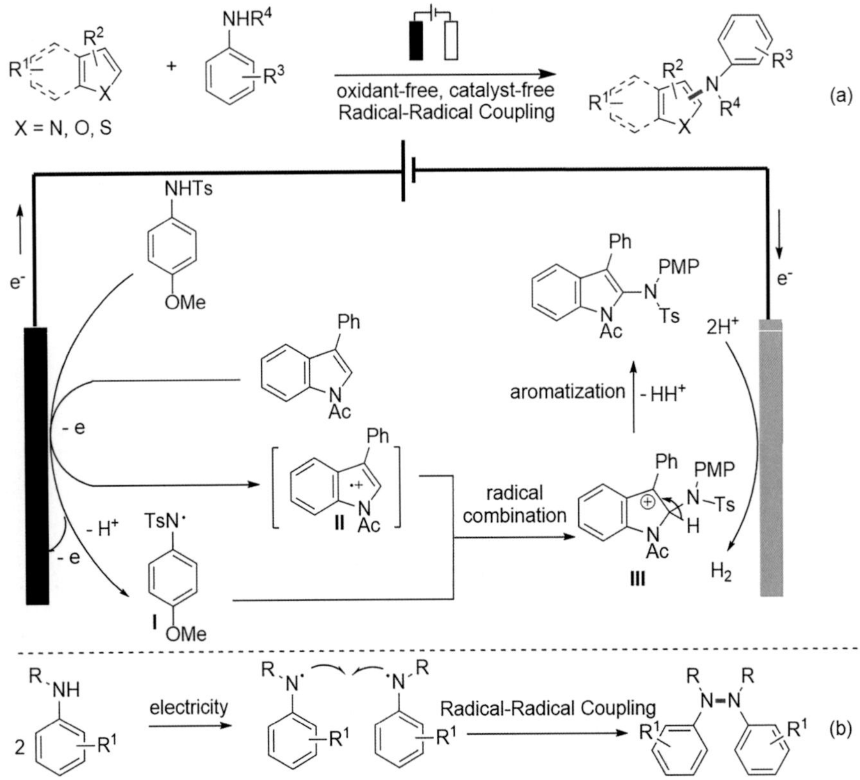

Scheme 2 Coupling reactions initiated by electro-oxidation to form *N*-radicals

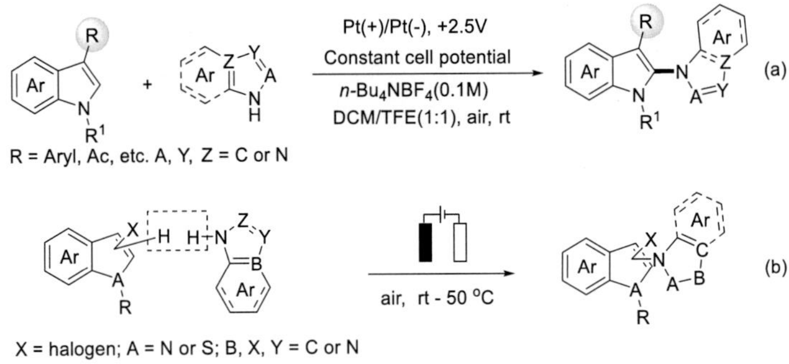

Scheme 3 Electro-oxidative coupling of azoles with indoles

retention of bromine was achieved. Dichloromethane played a crucial role in this reaction system; without it as a solvent, the reaction could not proceed smoothly. This method is not only applicable to pyrazole but also exhibits good compatibility with triazoles and tetrazoles. Furthermore, this method can be applied to the late-stage functionalization of bioactive molecules, demonstrating its potential in biologically relevant applications. Soon after, they successfully tested the preinstallation of bromine atoms at the C3 position of indole, with the voltage increased to 3.5 V, further exploring the dehydrogenative coupling reaction with pyrazole [8] (Scheme 3b). The substrate generality of this reaction remains highly broad, exhibiting excellent compatibility with various heterocyclic compounds.

Dehydrocoupling is much broader than a direct coupling to heterocycles. It is also an effective method for functionalizing heterocycles attached to hydrazones. Thus, electrochemical α-C – H phosphorylation was achieved with diphenylphosphine oxide in an attempt to use oxime as a free radical acceptor [9] (Scheme 4). The strategy provided expedient access to highly functionalized α-iminophosphine oxides with ample scope and broad functional group tolerance by means of mild, user-friendly electrolysis, in an undivided cell. For instance, aryl, alkynyl, and heterocyclic hydrazones were converted to the desired products in good yields. It is noteworthy that the reaction could also be conducted on a scale of 10 mmol, yielding the product in 80% yield. The aldehyde hydrazone substrate is initially oxidized to carbocation intermediate **I** under the electrochemical conditions, which then reacts with diphenylphosphine oxide to form the C – N bond, generating the aminyl radical intermediate **II**. Further anodic oxidation of **II** produces the cation **III**, followed by deprotonation to yield the imine product.

Subsequently, sulfoxonium ylides were functionalized by an electrochemical system using magnesium chloride as chlorine source, enabling an α-C – H chlorination reaction [10] (Scheme 5a). Furthermore, the authors observed the formation of *multi*-chlorinated products in the experiment, resulting in a certain decrease in selectivity. Upon combining α-benzoyl sulfoxonium ylides with diphenyl diselenide, it was found that the main product formed was the diselenide after the loss of DMSO, along with some *mono*-selenide products [11] (Scheme 5b). Under mild reaction conditions, the formation of transient diselenides, which are not stable, was achieved, and the reaction was unaffected by when the substrate contained drug substructures. In particular, a series of commercially available drugs such as naproxen, oxaprozin, isoxepac, probenecid, and dehydrocholic acid were all appropriate substrates for this electrochemical selenylation. Upon testing the obtained selenium-containing products for activity, it was found that some of them exhibited superior anti-inflammatory activity against *C. albicans* (MIC = 3.125–6.25 μg/mL), including *C. albicans* ATCC10231 and *C. albicans* ATCC98001. Additionally, by introducing 5 mmol of benzoic-acid-derived substrate, the gram-scale synthesis of α-keto *gem*-diselenide products was achieved in three steps.

2.2 *Csp^3^—H Dehydrogenative Coupling*

After successfully achieving Csp^2 – H dehydrogenative coupling reactions, the benzylic Csp^3–H azolation of 4-methoxyethylbenzene was also investigated,

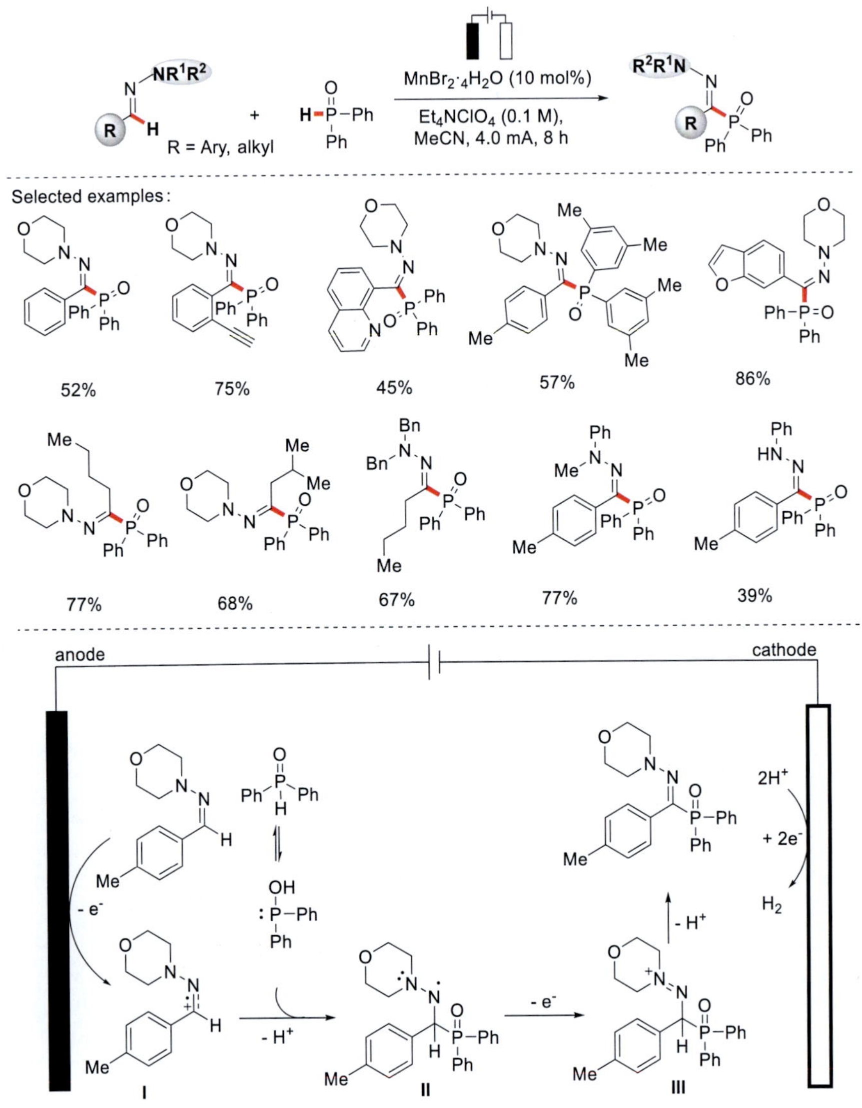

Scheme 4 Electrochemical oxidative phosphorylation of aldehyde hydrazones

which proceeded in an undivided cell under mild, catalyst- and chemical-oxidant-free reaction conditions [12] (Scheme 6). This strategy empowered the C–H azolation on primary, secondary, and even challenging tertiary benzylic positions selectively. The remarkable synthetic utility of this approach was highlighted by its easy scalability without overoxidation of products and ample scope with valuable functional groups. When substrates contain multiple potential reaction sites, it was observed that the azolation products formed regioselectively at *p*-alkoxy-activated

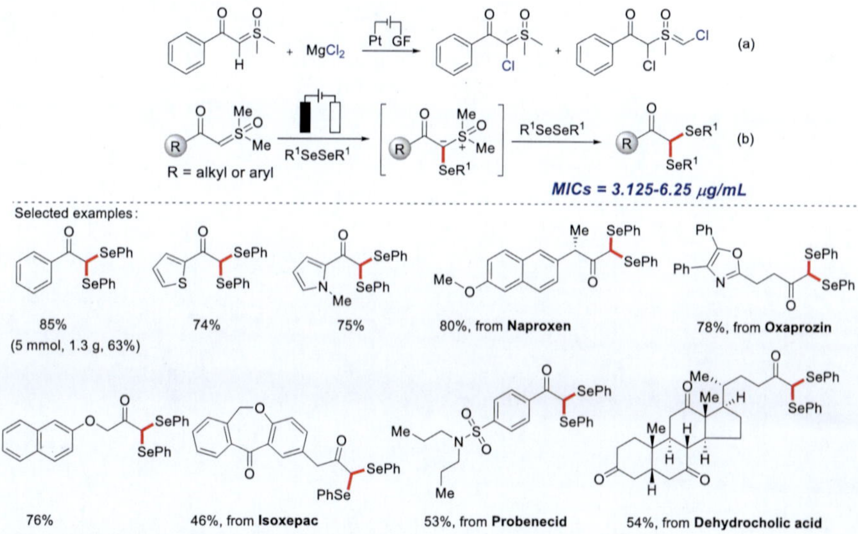

Scheme 5 Electrochemical selenylation of sulfoxonium ylides

benzylic positions. Moreover, the unique power of the versatile electro-oxidative C–H azolation approach set the stage for the efficient modification and assembly of highly functionalized bioactive compounds, such as sartan pharmaceuticals, providing a promising new platform to design new bond formation, sustainable oxidative functionalization reactions with broad practicality in synthetic and medicinal chemistry. Concurrently, a similar study by Xu et al. [13] focused mainly on electron-deficient benzylic positions and sulfonamides, achieving benzylic aminations. On the contrary, the research by Ruan et al. focuses on electron-rich benzylic azolations with similarities in the mechanism. This method can also be used for coupling two high-value complex reaction partners, albeit with decreased yields.

Apart from the benzylic C–H azolation reaction previously achieved, an alternative and easily accessible approach with a different C–H partner was developed [14] (Scheme 7). Ruan et al. selected α-carbonyl compounds for coupling reactions with tetrazoles, utilizing the electrolyte n-Bu_4NI to generate α-iodoketone intermediates in situ from ketones, which then undergo substitution reactions to yield the azolated products. The substrate scope of this reaction is extremely broad, allowing for azolation of various carbonyl compounds bearing aromatic rings, alkyl groups, or heterocycles in moderate to good yields. Importantly, azolation of the industrial solvent acetone is also effective. Furthermore, when methyl-ketone-containing drug molecules are chosen as substrates, selective azolation at the methyl position can also be achieved with high selectivity.

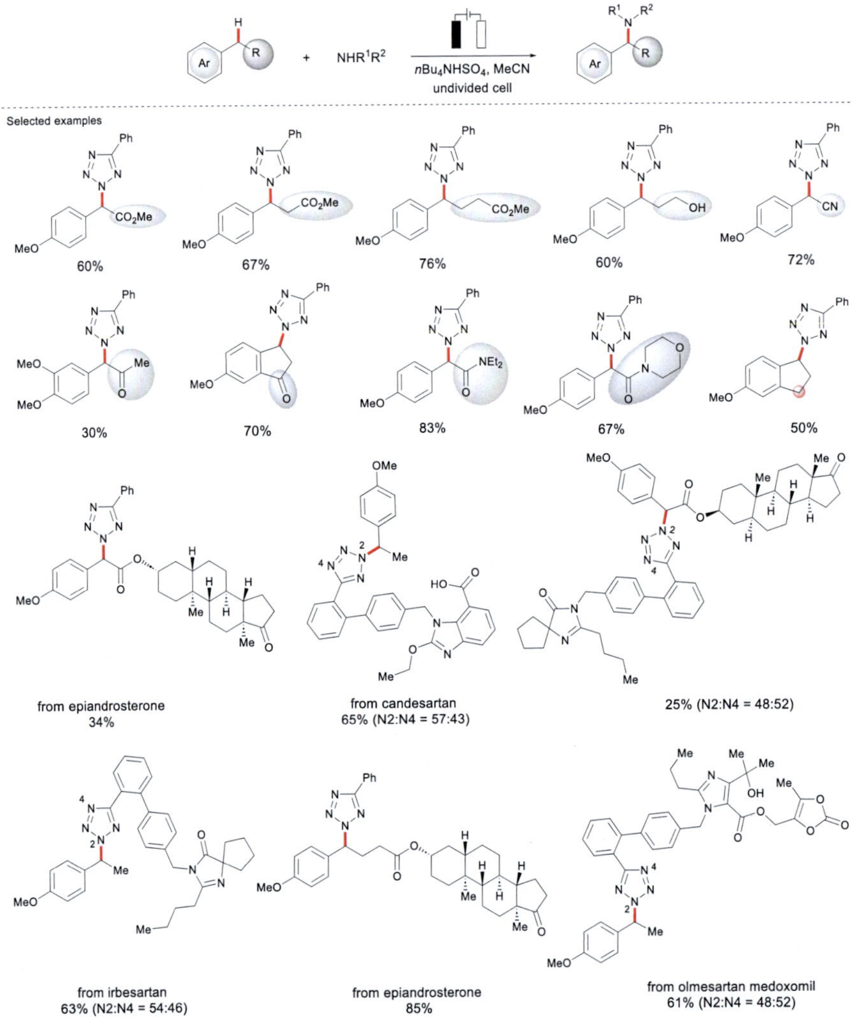

Scheme 6 Electro-oxidative azolation of benzylic C–H bonds

3 Heterocyclic Cyclization Reactions

3.1 Se-Substituted Heterocyclic Cyclization Reactions

Building upon the previously reported radical-mediated *N*-heterocyclic cyclization reactions, Ruan et al. aimed to further achieve electrochemical coupling cyclization reactions involving other heteroatomic elements, such as selenium [15] (Scheme 8). Based on the previous electrochemical intermolecular trifluoromethylation/radical cyclization of *N*-substituted acrylamides, a catalyst-free, metal-free, and

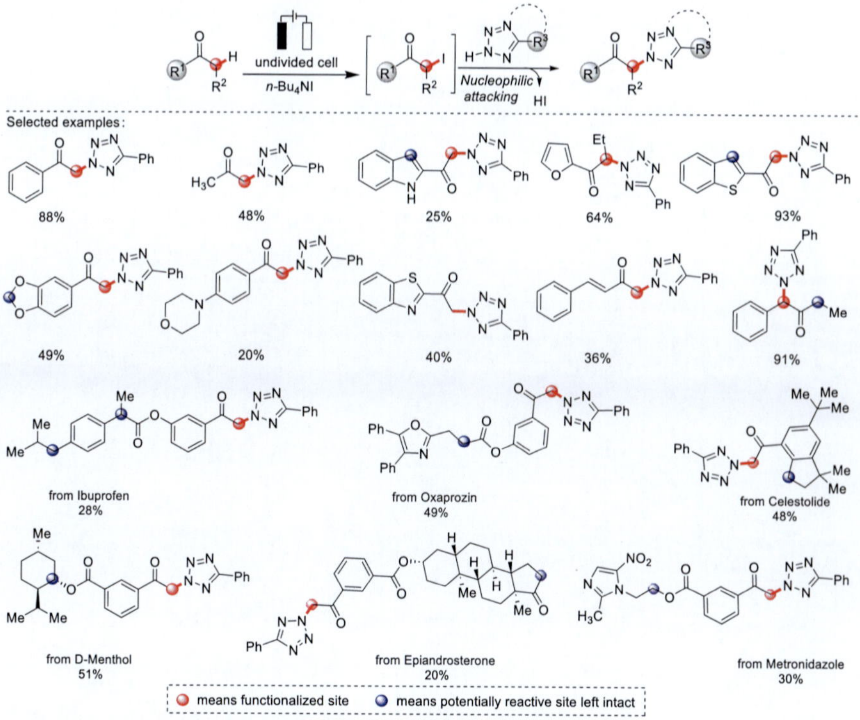

Scheme 7 Electrochemically enabled α-C(sp^3)–H azolation of ketones

environmentally friendly electrochemical selenylation/cyclization reaction of alkenes has been successfully developed with moderate to excellent yields. This selenium-incorporating transformation proceeds smoothly and tolerates a wide range of synthetically useful groups to deliver diverse functionalized benzheterocycles. Additionally, by appropriately modifying the structure of the alkene to incorporate a carboxylic acid or *N*-arylacrylamide, it was facile to obtain lactones and quinolinones. A variety of substituted *N*-arylcinnamamides were smoothly converted to the selenylated 3,4-dihydroquinolin-2(1*H*)-ones with high diastereoselectivity. The control experiments verified that aryl selenium chlorides could replace selenides in the reaction, wherein the aryl selenium cation directly participates, activating the olefin for carbonyl oxygen attack and cyclization. Furthermore, cyclic voltammetry experiments confirmed that diselenides are more easily oxidized compared to other substrates.

After modifying aryl alkenes to aryl alkynes, Ruan et al. discovered that in the presence of diaryl diselenides, it is easier to activate alkynes and further induce the neighboring heteroatoms to cyclize, yielding furan-type products [16] (Scheme 9a). This protocol is straightforward, and the chosen substrates are easily accessible, allowing for scalability. Through electrochemical synthesis, selenium-substituted furans can be synthesized with high chemoselectivity and excellent yields.

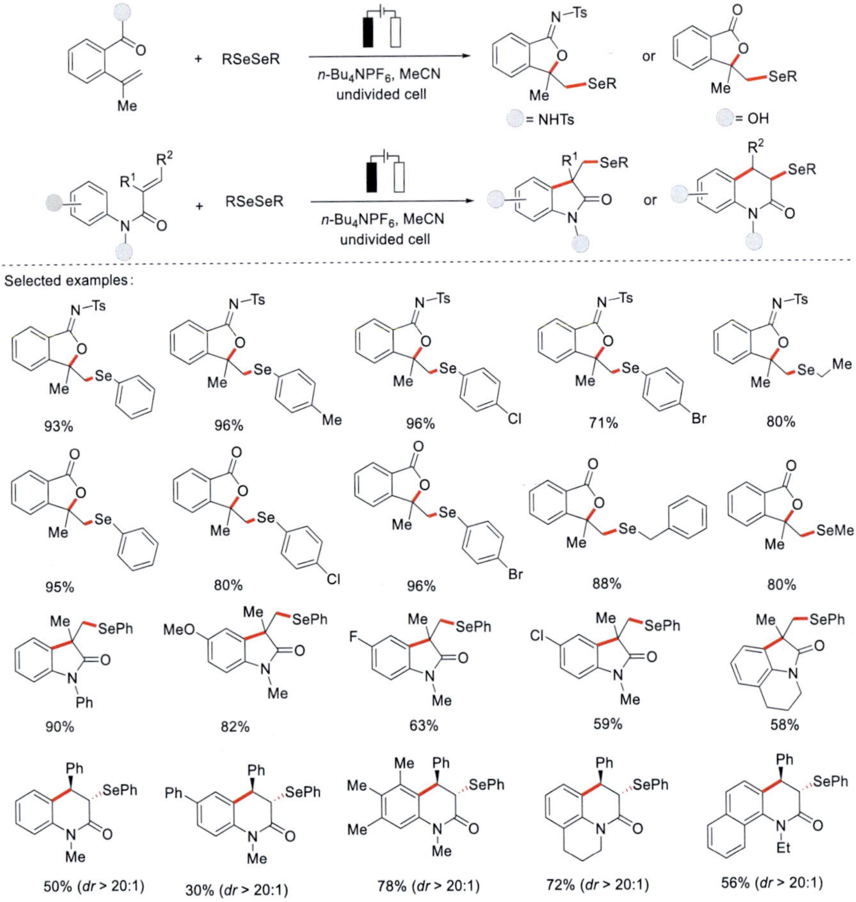

Scheme 8 Electrochemical selenylation/cyclization of alkenes

Moreover, a novel metal- and oxidant-free electro-oxidative selenylamination of *o*-aminophenacetylenes with diselenides for preparing 3-selenylindoles has also been developed with moderate to excellent yields [17] (Scheme 9b). Expanding the scope to include more reactive alkynyl aryl ketone partners, further applications can be found in the synthesis of 3-selenylated chromones [18] (Scheme 8c). Furthermore, using D_2O as a cosolvent allows for deuteration of 1,2-addition products of alkynes and selenides, enabling a highly selective synthesis of deutero-selenylated chalcones. Mechanistic experiments confirmed that the reaction was initiated by the anodic oxidation of diphenyl diselenide to form the selenium cation **A**, addition of **A** to the alkyne to generate intermediate **C**, cyclization to form oxyallyl cation **D**, and *O*-demethylation by selenium anions to provide the desired products and release PhSeMe, which was detected by GC-MS. At the same time, the alkynes are attacked by the aryl selenium anions produced in the system and then protonated to give the

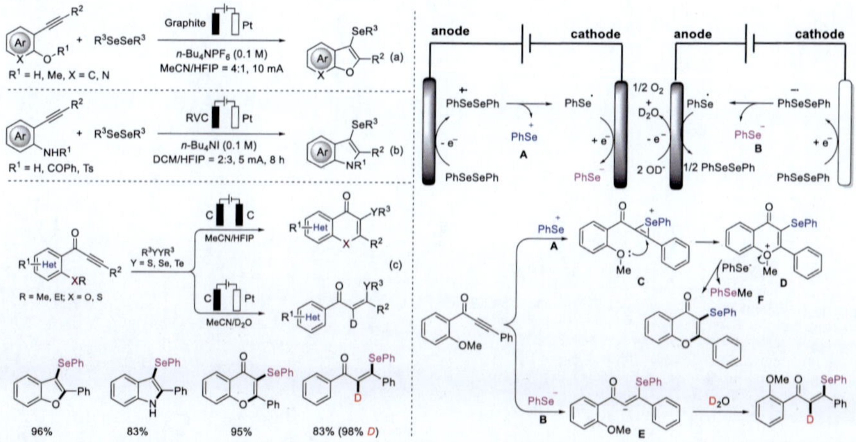

Scheme 9 Electrochemical selenylation/cyclization of alkynes

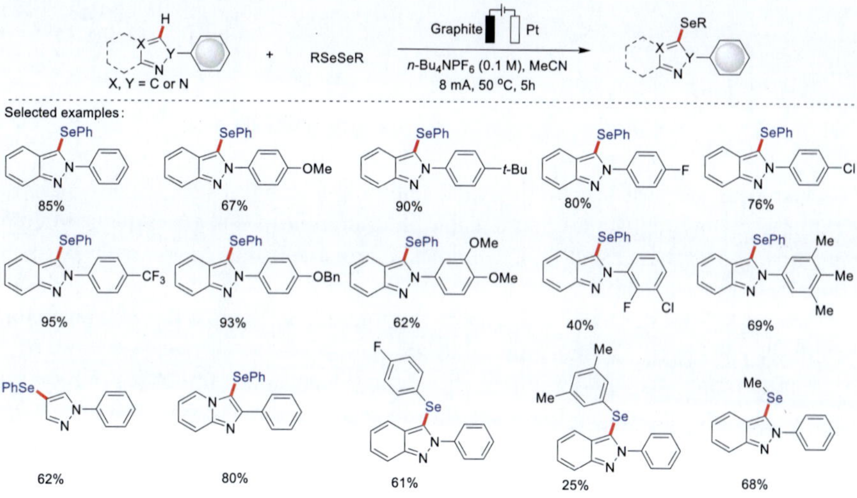

Scheme 10 Regioselective C–H selenylation of azoles

1,2-addition products of the alkynes. Finally, some of these compounds display anti-inflammatory activity, suggesting their potential as lead compounds for developing anti-inflammatory drugs.

Selenium-substituted heteroaromatic compounds have been proven to possess significant biological activities and serve as useful structural motifs. Based on previous functionalization and construction studies of selenium-containing heterocycles, Ruan et al. have successfully achieved the electrochemically dehydrogenative selenylation of imidazoles with various diselenides [19] (Scheme 10). The reaction does not require the addition of chemical oxidizing agents, and

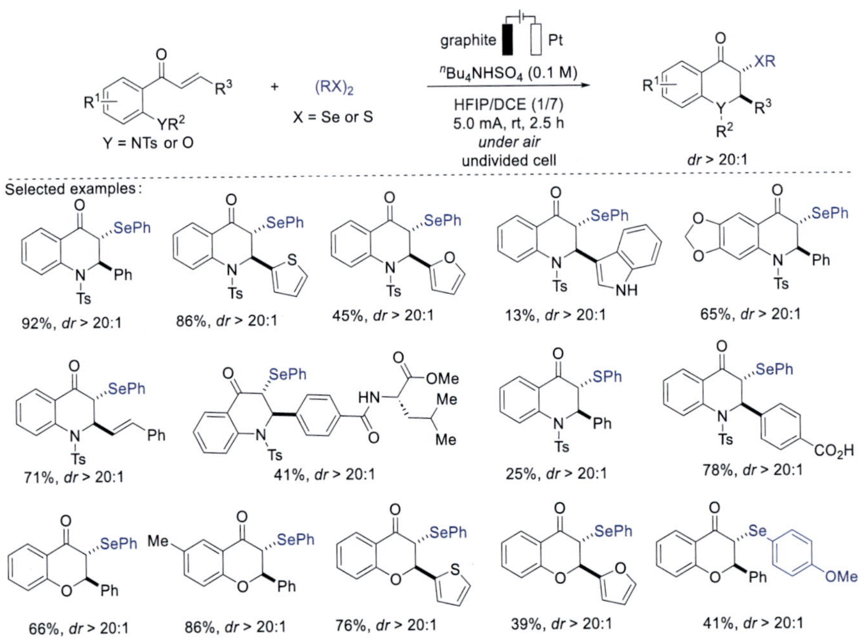

Scheme 11 Electrochemical selenylation of chalcones with diselenides

compared to the past use of elemental iodine as a catalyst [20], electrochemical synthesis is environmentally sustainable. Exploration of substrates with different functional groups revealed an exceptionally broad substrate scope with excellent functional group tolerance. Due to its simplicity of operation and easy accessibility of raw materials, this approach can be performed on gram-scale, demonstrating promising prospects for industrial synthesis.

Furthermore, the authors have developed a method for the synthesis of 3-selenylazaflavanones and 3-selenylflavanones via oxidant-free and metal-free electrochemical selenization reactions of chalcones with diselenides at room temperature [21] (Scheme 11). The method was carried out under mild conditions and enabled the realization of not only azaflavones but also oxaflavone products. The method demonstrates broad applicability across a diverse range of substrates and exhibits remarkable regioselectivity and stereoselectivity toward substrates featuring highly reactive functional groups, such as indolyl, styryl, furyl, and thiophenyl moieties. Furthermore, the selenylated products were obtained in yields ranging from moderate to excellent. In addition, the reaction can also be easily scaled up with high efficiency. Detailed mechanistic studies, including controlled experiments, revealed that selenyl radicals and selenium cations may be simultaneously involved in this electrochemical process.

3.2 *Other Heterocyclic Cyclization Reactions*

In addition to the oxidative dehydrogenative coupling reactions discussed above, radical addition reactions with olefins have been demonstrated to possess significant capabilities. By utilizing electro-oxidation to generate radicals, further cyclization reactions occur with activated olefins to form cyclic products. While there are numerous radical-involved cyclization reactions in the fields of metal- or photo-catalysis, such as the advancement of selective catalytic reactions for radical cyclization, employing electrocatalysis enhances the chemical toolbox for the synthesis of enantioenriched specialty chemicals. Compared to previously reported reactions, Ruan et al. have avoided the use of excess oxidants or metals, thus achieving serial reactions of CF_3/CF_2H radicals under environmentally favorable conditions [22] (Scheme 12). The proposed mechanism of the reaction involves the fluoroalkyl radical source undergoing anodic oxidation, leading to the removal of SO_2 and the formation of a radical species. This radical is then captured by the olefin of the arylacrylamide, resulting in the generation of alkyl radical **III**. These alkyl radicals further attack the aromatic ring to promote cyclization, eventually yielding products through oxidation and aromatization. The reaction is convenient to operate at room temperature and exhibits good chemoselectivity. In most cases, *N*-arylacrylamide derivatives are used, providing products containing synthetically useful trifluoromethylated oxindoles. Furthermore, the electrochemical fluoroalkylation/cyclization regime was not restricted to *N*-arylacrylamides. Indeed, *N*-arylcinnamamides were also successfully converted into fluoroalkylated 3,4-dihydroquinolin-2(1*H*)-ones with high diastereoselectivity (*dr* > 20:1) by slightly modified reaction, such as replacing the electrolyte with Bu_4NPF_6.

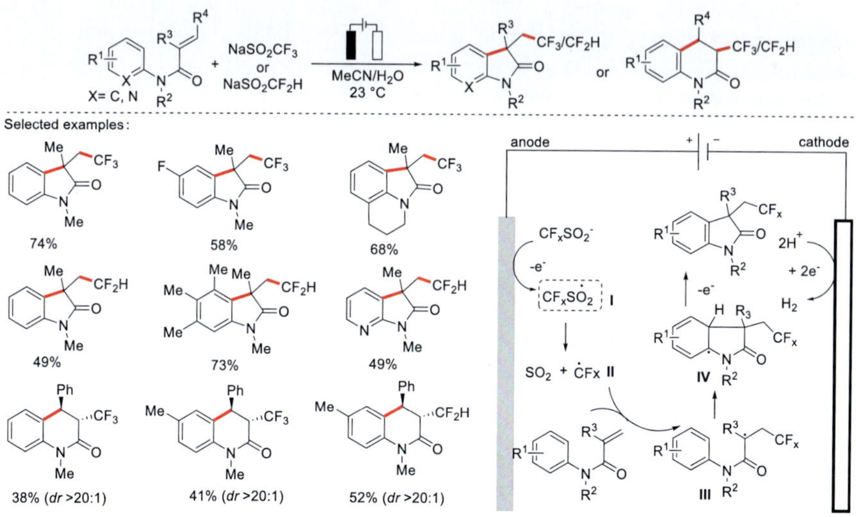

Scheme 12 Direct electrochemical tri- and difluoroalkylation/cyclization

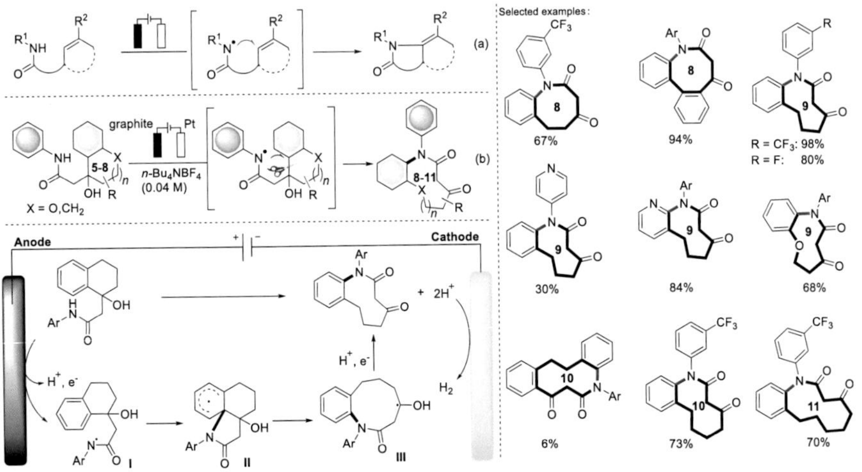

Scheme 13 Direct electrochemical synthesis of annulated medium-sized lactams

Aside from the ring formation involving fluoroalkyl radicals, employing amidyl-radical-induced ring formation to fabricate heterocyclic compounds holds significant importance in synthetic chemistry. Xu [23] and Moeller [24] reported the synthesis of highly substituted benzimidazolone derivatives through a dehydrogenative coupling cascade. This process utilizes an intramolecular reaction of *N*-arylamidyl radicals with alkynes via 5-exo-dig cyclization or 6-endo-trig cyclization, ultimately leading to the formation of 5- or 6-membered ring products (Scheme 13a). Inspired by these reactions, arylamidyl-radical-induced ring-opening by *β*-C–C bond cleavage and concomitant cyclization was successfully achieved by preinstalling a six-membered ring bearing a tertiary alcohol in the substrates, through a unique remote amidyl migration under mild, metal- and external-oxidant-free conditions in a simple undivided cell. This strategy exhibits a broad substrate scope, affording synthetically challenging medium-sized lactams, such as (hetero)aryl-fused 8-11-membered ring and heteroatom-tethered rings, with high yields and excellent functional group tolerance [25] (Scheme 13b). Electrochemical formation of arylamidyl radicals under mild conditions without the need for catalysts and oxidants, as opposed to transition metal-catalyzed addition with stoichiometric amounts of oxidizing agents, offers significant advantages. Moreover, the reaction mechanism was further verified, confirming the formation of arylamidyl radicals through control experiments and DFT calculations, and finding that the pathway involving radical addition of another aryl ring via a 5-membered ring transition state and subsequent C–C bond cleavage is most suitable.

The authors have also extended their work to dehydrocyclization. An electrochemical thiocyanation/cyclization reaction of aldehyde-derived hydrazones was developed in the absence of an external oxidant and catalyst [26] (Scheme 14). Unlike previous thiocyanation reactions, this electrosynthetic method enables the cascade C-H thiocyanation/cyclization to be achieved by mild, direct electrolysis in

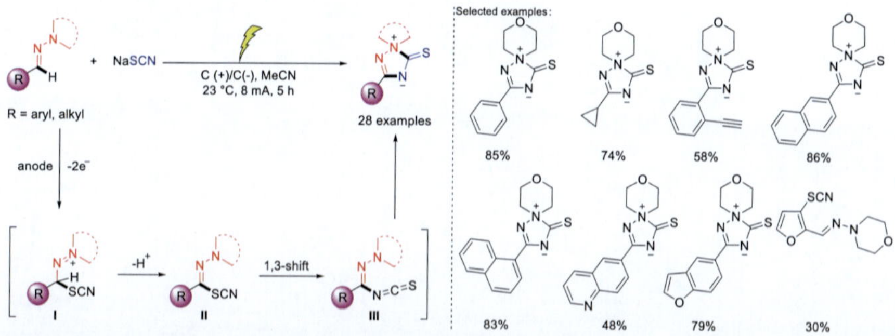

Scheme 14 Electrochemical synthesis of sulfur-containing triazolium inner salts

an undivided cell without the addition of halogens and stoichiometric oxidants. In this protocol, commercially available and inexpensive sodium thiocyanate was used, which served as a dual role of thiocyanate reagent and electrolyte. The strategy provides easy access to functionalized sulfur-containing triazolium inner salts with a sufficient range of different functional groups, including strained cyclopropyl, terminal alkynes, naphthalene rings, quinolines, or furans. Based on the results of the mechanistic studies, Ruan et al. proposed the following reaction mechanism: the aldehyde hydrazone loses electrons at the anode, promoting nucleophilic attack by the anion of thiocyanate. After a 1,3-shift isomerization of thiocyanic acid to form isothiocyanic acid, finally the product is obtained by cyclization.

After establishing the methodology for electrochemical dehydrogenative coupling, a synthetic application aimed at further developing and utilizing these methods in peptide-based drug molecules has been further developed. An initial experiment involved the synthesis of 2-vinylsulfonamides, wherein the addition of selenide as catalysts under electrochemical conditions facilitated directed dehydrogenative coupling to synthesize indoles. When applying this method to multiple peptide chains, they consistently obtained high yields of cyclized indole-containing peptide products. This provides a promising pathway for the specific synthesis of peptides containing indole moieties [27] (Scheme 15).

The preceding discussion has provided a concise overview of the advancements in electrosynthesis concerning N- and Se-substituted heterocyclic compounds. These compounds have emerged as pivotal areas in the realm of new drug development, owing to their diverse molecular structures and broad spectrum of biological activities. Moreover, they offer distinct advantages such as heightened efficiency, reduced toxicity, and eco-friendliness. Electrosynthesis, with its sustainable redox properties, presents a superior alternative to conventional chemical synthesis methods within the industry. Notably, it enables multifaceted electrocatalytic modifications of peptide compounds, offering cost-effectiveness, efficiency, and stability. It is anticipated that the near future will witness the emergence of novel peptide heterocyclic drug structures with enhanced activity, propelled by these advancements.

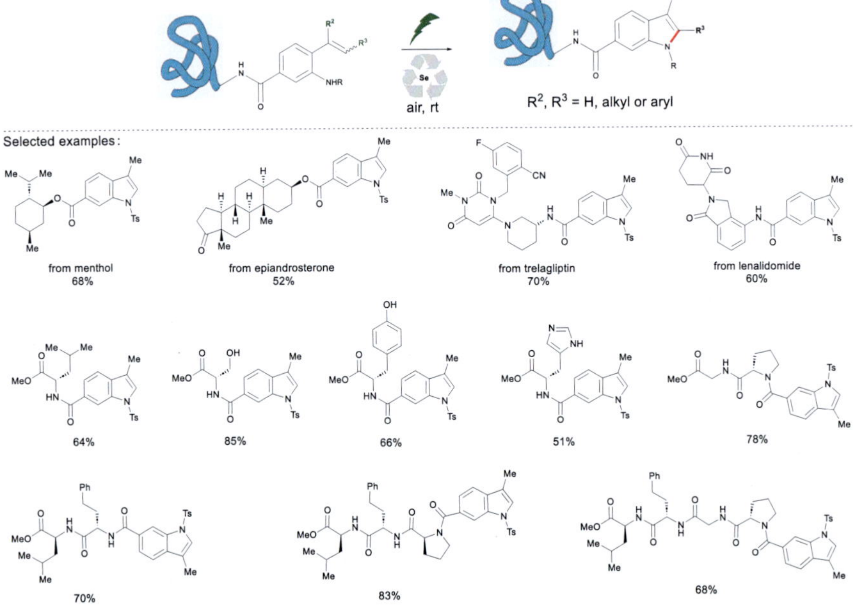

Scheme 15 Electrocatalytic cyclization of 2-vinylanilides toward indoles of peptide labeling

References

1. Pratley C, Fenner S, Murphy JA (2022) Chem Rev 122:8181–8260
2. Yamamoto K, Kuriyama M, Onomura O (2021) Chem Rec 21:2239–2253
3. Davies J, Svejstrup TD, Fernandez Reina D, Sheikh NS, Leonori D (2016) J Am Chem Soc 138: 8092–8095
4. Zhang Y, Lin Z, Ackermann L (2021) Chem A Eur J 27:242–246
5. Peng X, Zhao J, Ma G, Wu Y, Hu S, Ruan Z, Feng P (2021) Green Chem 23:8853–8858
6. Gieshoff T, Kehl A, Schollmeyer D, Moeller KD, Waldvogel SR (2017) J Am Chem Soc 139: 12317–12324
7. Zhou N, Zhao J, Sun C, Lai Y, Ruan Z, Feng P (2021) J Org Chem 86:16059–16067
8. Zhou N, Yu J, Hou L, Wu X, Ruan Z, Feng P (2022) Adv Synth Catal 364:35–40
9. Xu Z, Li Y, Mo G, Zheng Y, Zeng S, Sun PH, Ruan Z (2020) Org Lett 22:4016–4020
10. Zhang L-M, Fu Z-H, Yuan D-F, Guo M-Z, Li M, Wen L-R, Zhang L-B (2023) Tetrahedron Lett 114:154244
11. Xu Z, Yao J, Zhong K, Lin S, Hu X, Ruan Z (2023) J Org Chem 88:5572–5585
12. Ruan Z, Huang Z, Xu Z, Zeng S, Feng P, Sun P-H (2021) Sci China Chem 64:800–807
13. Hou ZW, Liu DJ, Xiong P, Lai XL, Song J, Xu HC (2021) Angew Chem Int Ed 60:2943–2947
14. Fang S, Zhong K, Zeng S, Hu X, Sun P, Ruan Z (2023) Chem Commun 59:11425–11428
15. Cheng X, Hasimujiang B, Xu Z, Cai H, Chen G, Mo G, Ruan Z (2021) J Org Chem 86:16045–16058
16. Hasimujiang B, Lin S, Zheng C, Zeng Y, Ruan Z (2022) Molecules 27:6314
17. Hasimujiang B, Zeng Y, Zou S, Zhong K, Su L, Hu X, Ruan Z (2024) Isr J Chem 64: e202300088
18. Hasimujiang B, Zhu J, Xu W, Wang H, Hu X, Ruan Z (2023) Adv Synth Catal 365:2929–2935
19. Lin S, Cheng X, Hasimujiang B, Xu Z, Li F, Ruan Z (2021) Org Biomol Chem 20:117–121

20. Dey A, Hajra A (2019) J Org Chem 84:14904–14910
21. Zeng S, Zeng Y, Wang H, Sun P, Ruan Z (2024) J Org Chem 89:4074–4084
22. Ruan Z, Huang Z, Xu Z, Mo G, Tian X, Yu XY, Ackermann L (2019) Org Lett 21:1237–1240
23. Xu F, Long H, Song J, Xu H-C (2019) Angew Chem Int Ed 58:9017–9021
24. Xu H-C, Campbell JM, Moeller KD (2014) J Org Chem 79:379–391
25. Xu Z, Huang Z, Li Y, Kuniyil R, Zhang C, Ackermann L, Ruan Z (2020) Green Chem 22:1099–1104
26. Li Y, Huang Z, Mo G, Jiang W, Zheng C, Feng P, Ruan Z (2021) Chin J Chem 39:942–946
27. Zeng S, Fang S, Cai H, Wang D, Liu W, Hu X, Sun P, Ruan Z (2022) Chem Asian J 17: e202200762

Top Heterocycl Chem (2024) 61: 55–90
https://doi.org/10.1007/7081_2024_80

Published online: 27 September 2024

Electrochemical Heterocyclic Ring-Formation Reactions by Making C–N and N–N Bonds

Alexander Sandvoß, Johannes Winter, Tobias Prenzel, and Siegfried R. Waldvogel

Contents

Abstract Nitrogen heterocycles are widely found as structural motifs in pharmaceuticals, active pharmaceutical ingredients (APIs), natural products, reagents, or other high value-added products. Electrochemistry as a modern synthesis technique, considering in particular its sustainable and environmentally friendly aspects, has enhanced the toolbox of organic chemists. Electrochemical [3 + 2]-cycloadditions, cyclization of electrochemically generated amidyl radicals and analogous moieties,

Alexander Sandvoß, Johannes Winter, and Tobias Prenzel contributed equally to this work.

A. Sandvoß, J. Winter, and T. Prenzel
Department of Chemistry, Johannes Gutenberg University, Mainz, Germany

S. R. Waldvogel (✉)
Department of Chemistry, Johannes Gutenberg University, Mainz, Germany

Institute of Biological and Chemical Systems-Functional Molecular Systems (IBCS-FMS), Karlsruher Institut für Technologie, Karlsruhe, Germany

Max-Planck-Institute for Chemical Energy Conversion, Mülheim an der Ruhr, Germany
e-mail: siegfried.waldvogel@cec.mpg.de

and cathodic nitro reduction of nitroarenes with subsequent cyclo-condensation are reactivities in heterocycle synthesis proven to have synthetic potential through their application in several reported electro-organic protocols.

Keywords Cycloaddition · Electrochemistry · N-Heterocycles · N-N Coupling · Nitro reduction

1 Introduction

The pursuit of more sustainability has led to a renaissance of electrosynthesis and the development of efficient modern synthetic strategies toward high value-added products [1–3]. The avoidance of reagent waste and simplifying downstream processing represent important features for later translation into various applications [4–6]. Synthetic approaches toward *N*-heterocycles are especially important for organic chemistry, as they are a widely found structural motif found in natural products. Since they are capable of hydrogen bonding, they are also important structural motifs in active pharmaceutical ingredients (APIs) and other high-value fine chemicals [7–11]. The electrochemical conversion of arenes has been a long-standing concept, whereas their electrochemical synthesis evolved vividly within the past decade. In recent years, the electrochemical synthesis of heterocyclic compounds has been dealt with in several overviews with different emphases, to which we would like to refer at this point [12–19].

In principle, electrochemical syntheses can be divided into two alternative approaches [20–23]. Either the substrate is directly reacted at the electrodes, or a redox mediator transfers the electric charge from the electrodes to the substrate, triggering the desired redox transformation (Scheme 1). The direct electron transfer can take place on an inert or an active electrode material. Most commonly, noble metals, such as platinum, or carbon allotropes, like graphite (C_{gr}) glassy carbon (GC) and carbon felt (C_{felt}), or high-performance materials, such as boron-doped diamond (BDD) and reticulated vitreous carbon (RVC), are used as inert electrodes [24, 25]. When a layer of immobilized high-valent metal species is formed on the electrode surface, the electrode is considered as active anode, for example, NiOOH

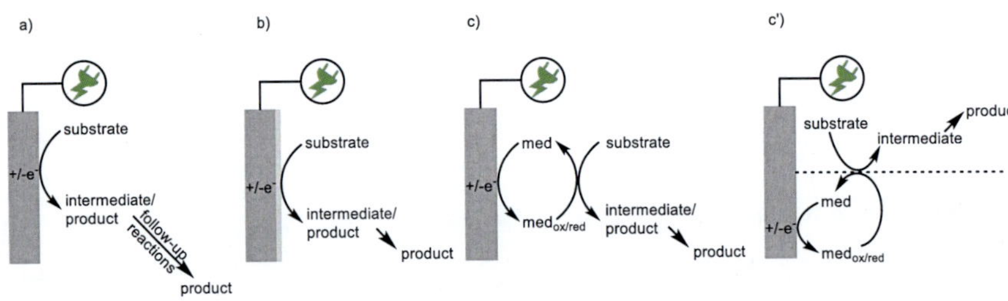

Scheme 1 General modes of electron transfer for electro-organic synthesis

and others [26, 27]. This redox-active layer, which is regenerated by the electric current, catalyzes the electron transfer, therefore often improving the selectivity of the overall transformation. This concept is taken to the extreme in mediated electrolysis when a redox mediator is employed as a soluble redox shuttle from the electrode to the substrate. Thereby, even biphasic setups can be realized. Halides are frequently used as mediators but also as supporting electrolytes, rendering them a dual role as reagent as well [28, 29]. Noteworthily in this regard is the anodic oxidation of iodide, which could lead to a plethora of oxidized species, including iodine radicals, elemental iodine, or iodonium cations. Depending on the electrolysis conditions, these and other oxidized species with even higher oxidation states are active species in electrochemical reactions [29, 30]. Consequently, utilizations of hypervalent iodine (III) reagents have been established as powerful-mediated synthesis strategies [31, 32].

In all these approaches, the electrode material is of great significance for the outcome and the performance of the electro-organic reaction. Not only the composition of the material but also its surface and the modifications thereof play a decisive role for the electron transfer and subsequent reactions [33]. Noteworthily, the action of the electrode cannot be separately considered from the electrolyte that usually functions as a couple. As a consequence, the composition of the electrolyte has a huge impact on the electrochemical reactivity [34]. Similarly, the choice of solvent and further additives has to be taken into account carefully [35–38]. Whereas early electro-organic syntheses frequently made use of toxic heavy metal electrodes, such as mercury pool electrodes, modern approaches increasingly rely on the use of high-performance materials, such as boron-doped diamond [39]. The efficiency of the reaction is often improved by circumventing various negative effects, such as cathodic corrosion of the electrode material [40, 41]. Furthermore, to conduct a well-working preparative electrolytic process, any counter reactions should not be neglected [42].

2 Electrochemically Induced [3 + 2] Cycloaddition in the Synthesis of Heterocycles

2.1 [3 + 2]-Cycloaddition in the Synthesis of Heterocycles

Pericyclic reactions are a powerful technique in the synthesis of cyclic molecules, including heterocycles. Particularly noteworthy are 1,3-dipolar cycloadditions, which have played an important role in the access to heterocycles ever since the first reports in the 1960s by *R. Huisgen* [43, 44]. In those reactions, a 1,3-dipolar species engages with a so-called dipolarophile in a cycloaddition reaction. Especially, nitrile oxides, azides, or nitrile imines have been found frequently used as the 1,3-dipolar components, whereas exemplary dipolarophiles are alkenes, alkynes, enols, or imines (Scheme 2). In recent years, electro-organic synthesis immensely

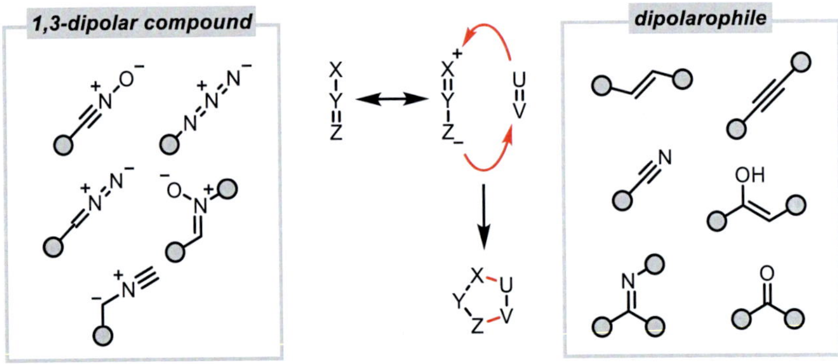

Scheme 2 General overview of [3 + 2]-cycloadditions

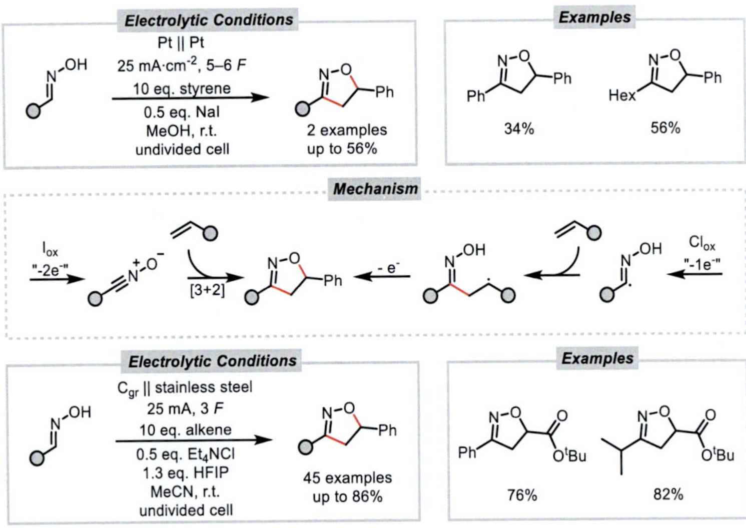

Scheme 3 Electrochemical synthesis of isoxazolines by halide-mediated anodic oxidation of aldoximes and subsequent [3 + 2]-cycloaddition [45, 46]

contributed to this field, as it offers an elegant and effective way for the generation and activation of both 1,3-dipolar compounds and the respective dipolarophiles, while avoiding the use of harmful oxidizers.

Isoxazoles and isoxazolines are widely accessible by [3 + 2]-cycloaddition of nitrile oxides with alkenes or alkynes. The electro-organic synthesis of these structural motifs has been reported in several protocols by oxidation of aldoximes to generate nitrile oxides as 1,3-dipolar compounds. These methods can be differentiated into direct or halide-mediated oxidation of aldoximes to nitrile oxides. The first electrochemical synthesis was reported by *Shono* et al. in 1989 (Scheme 3, top) [45]. Here, an electrochemical protocol with a domino oxidation-reduction sequence of aldoximes to nitriles was investigated. To demonstrate the mechanism of iodide-

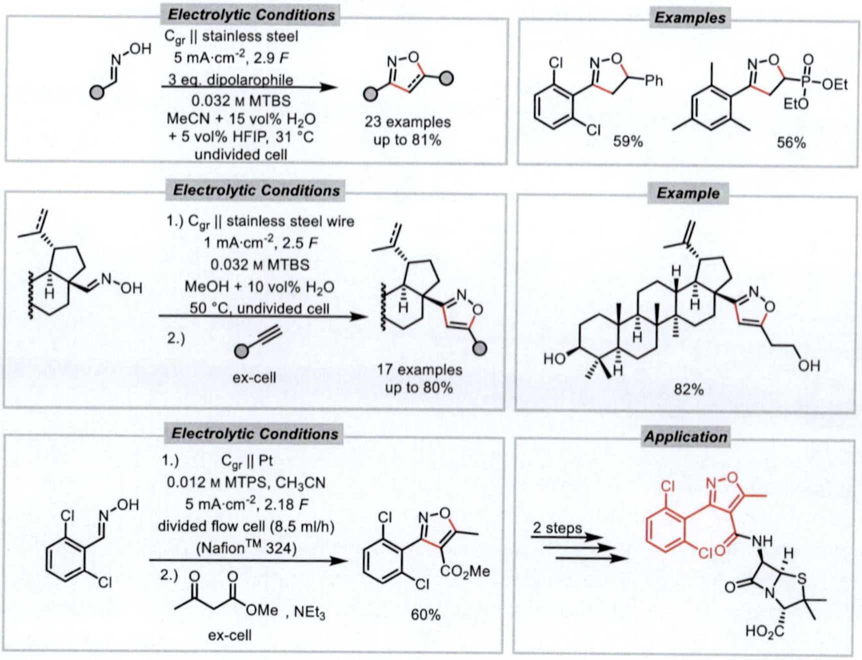

Scheme 4 Electrochemical synthesis of isoxazoles and isoxazolines by direct anodic oxidation of aldoximes and subsequent [3 + 2]-cycloaddition [48–50]

mediated anodic oxidation of an aldoxime to a nitrile oxide before its eventual reduction to a nitrile, electrolysis with styrene was performed to trap the nitrile oxide by a [3 + 2]-cycloaddition. The synthetic potential for electrochemical access to isoxazoles and isoxazolines and the electrolytic conditions were not further investigated at that time. Later, *Reid* et al. reported an electrochemical protocol by chloride-mediated oxidation with tetraethylammonium chloride in an anhydrous medium (Scheme 3, bottom). HFIP was utilized as additive to stabilize radical cations. A wide range of aryl and alkyl aldoximes could be converted. In mechanistic investigations by kinetic (time course) modeling and density functional theory, however, the initially proposed 1,3-dipolar cycloaddition could not be confirmed. Therefore, a stepwise, radical-mediated mechanism was proposed, but it could not be fully confirmed either [46].

In contrast, *Waldvogel* et al. reported the generation of nitrile oxides by direct oxidation of aldoximes at graphite anodes [47]. With this approach, the extensive platinum corrosion in the electrochemical protocol of *Shono* et al. could be avoided. For the domino oxidation-reduction sequence from aldoximes to nitriles, higher yields were obtained and the electrolysis could be conducted in flow [48]. In addition, a broadly applicable electrochemical protocol to isoxazoles as well as isoxazolines was developed (Scheme 4, top). Functional-group tolerance as well as a broad variety of alkenes and alkynes could be achieved using various in-cell or

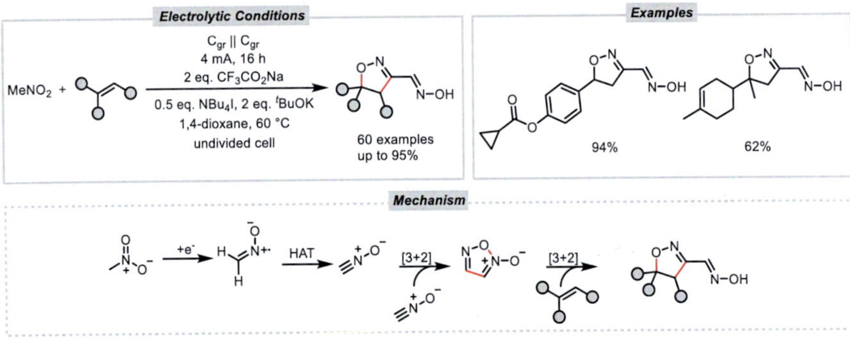

Scheme 5 Electrochemical synthesis of isoxazolin-3-aldoximes by activation of nitromethane and double [3 + 2]-cycloaddition [51]

ex-cell approaches. 2,6-Disubstituted aldoximes showed the best performance by preventing self-cycloaddition. Furthermore, by using a nitrile as dipolarophile, the corresponding 1,2,4-oxadiazole was obtained as a proof of concept. Recyclability of solvents, especially reuse of HFIP, could be demonstrated by usage in repeated electrolysis after distillation obtaining constant performance [49]. Furthermore, *Waldvogel*, *Shults*, *Turks*, and coworkers developed the synthesis of novel triterpenoid-isoxazole conjugates by electrosynthesis of betulin-derived nitrile oxides (Scheme 4, middle). Due to the rigid aliphatic backbone as well as the strong steric protection, stable nitrile oxides could be isolated, which allow ex-cell [3 + 2]-cycloaddition by adding the dipolarophile after electrolysis [50]. In addition, *Waldvogel* et al. reported a flow protocol for the synthesis of an isoxazole as building block for a penicillin precursor of dicloxacillin by cycloaddition of nitrile oxides and β-ketoesters (Scheme 4, bottom). Direct anodic oxidation is performed in a divided electrochemical flow cell with Nafion™ as separator material to suppress nitrile oxide reduction. The 1,3-dipolar cycloaddition is performed ex-cell with the enol tautomer of an 1,3-dicarbonyl. Subsequently, the isoxazole is obtained after water elimination and aromatization. The key nitrile oxide intermediate could be identified by GC-MS [48].

Access to isoxazolin-3-aldoximes has been reported by *Zhang* et al. by electrochemical activation of nitromethane and double [3 + 2]-cycloaddition (Scheme 5). Activation of nitromethane involves its conversion to fulminic acid (HCNO). Therefore, nitromethane is first reduced to an *N*-oxy imidyl radical at a graphite cathode by deoxygenation and converted to fulminic acid via HAT by an oxidized iodine species or a trifluoromethyl radical, both generated at the anode. Subsequently, [3 + 2]-cycloaddition with itself occurs to yield 1,2,5-oxadiazole-*N*-oxide, which has been detected via GC-MS. In addition, the dimer acts as 1,3-dipolar compound and undergoes a second [3 + 2]-cycloaddition with alkenes to enable access to isoxazolin-3-aldoximes. A broad scope of alkenes with multifunctional group tolerance could be converted, demonstrating the applicability of the electro-organic reaction [51].

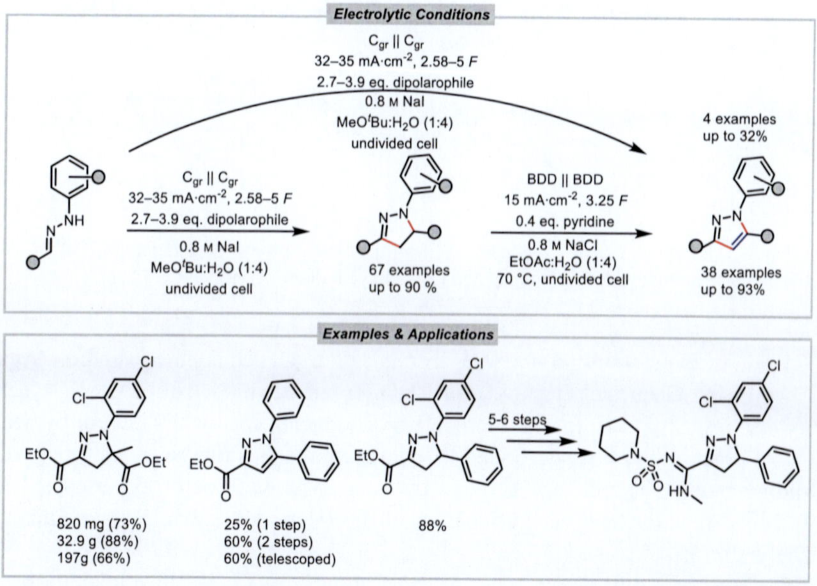

Scheme 6 Electrochemical synthesis of pyrazolines and pyrazoles by halide-mediated oxidation of hydrazones and electrochemical aromatization of pyrazolines to pyrazoles [52–54]

In 2022, *Waldvogel* et al. reported an electrochemical synthesis of pyrazolines and pyrazoles by iodide-mediated anodic oxidation, which could be achieved by biphasic system using sodium iodide as an inexpensive mediator and readily available supporting electrolyte (Scheme 6). The electrochemical reaction takes place in the aqueous layer, which prevents side reactions of the hydrazone and the dipolarophile. In addition, the robustness of the electrochemical protocol based on the biphasic electrolyte is demonstrated by a large scope. Excellent functional group tolerance was achieved, which was underscored by access to important bioactive compounds and building blocks. For example, an important intermediate for a class of drugs used as CB_1 receptor antagonists was obtained in good yields [52]. The scalability of this reaction was investigated and further optimized with focus on the herbicide safener mefenpyr-diethyl. In this study, the electrochemical reaction could even be performed on a 197-g scale. A strong influence of the applied electrolytic conditions for *cis*- and *trans*-isomers of the hydrazones was observed. Here, the reaction of the (*Z*)-hydrazone enables the omission of an additional organic solvent, whereby the two-phase electrolyte system is obtained solely by the dipolarophile and the aqueous sodium iodide solution [53].

Mechanistically, iodide oxidation takes place in the aqueous layer, followed by oxidation of the hydrazone by the oxidized iodine species at the interface. The latter oxidation generates nitrile imines as the 1,3-dipolar compounds, enabling [3 + 2] cycloadditions with alkenes to give access to pyrazolines. In this study, alkynes could also be converted, but the yields were limited due to slower cycloadditions [52]. Therefore, the same lab developed a further electrochemical protocol for the

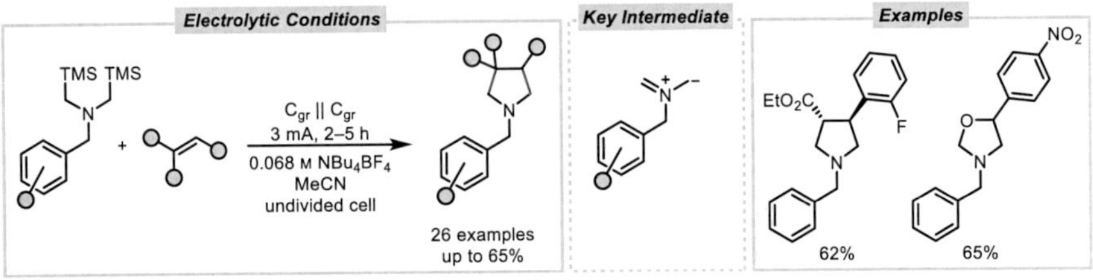

Scheme 7 Electrochemical synthesis of pyrrolidines by direct oxidation of *N,N*-bis (trimethylsilylmethylene)benzylamines and [3 + 2]-cycloaddition with alkenes [61]

oxidation of pyrazolines to pyrazoles by chloride-mediated anodic oxidation. Again, carbon-based electrodes, in this case BDD, could be utilized in a simple electrochemical protocol [54]. Furthermore, electrochemical functionalization of pyrazoles in the 4-position has been described in several approaches for the synthesis of useful building blocks. The electrochemical two-step synthesis could be performed in a telescoped reaction without purification of the intermediate while maintaining the same yield [55–60].

The electrochemical synthesis of pyrrolidines by [3 + 2]-cycloaddition was described by *Banerjee* et al. from *N,N*-bis(trimethylsilylmethylene)benzylamines and alkenes (Scheme 7). Direct oxidation of the bis-silyl component generates azomethine ylides as 1,3-dipolar compounds, which undergo [3 + 2]-cycloaddition with alkenes. In addition, aldehydes as dipolarophiles could be converted into oxazolidines. Sustainable and environmentally benign electrolytic conditions make this electrochemical protocol promising for the generation and further use of nonstabilized azomethine ylides [61].

In addition to the electrochemical generation of 1,3-dipolar compounds, the electrochemical activation of dipolarophiles has been reported. An electrochemical synthesis of 1,3-oxazoles from carboxylic acids and isocyanides was reported by *Xia* et al. (Scheme 8). Here, the carboxylic acid was activated by iodide-mediated deoxygenation with triphenylphosphine, resulting in an acyloxyphosphonium ion as dipolarophile. This method enables the preparation of various oxazoles with high functional group tolerance under simple and convenient electrolytic conditions. Numerous synthetic applications have been demonstrated in a broad scope highlighted by late-stage functionalization of several biologically active compounds, like ibuprofen zaltoprofen, and etodolac [62].

Electrochemical synthesis of 4-acyl-1,2,3-triazoles was established by *Bera* et al. from propargyl alcohols and sodium azide (Scheme 9). Initially, the alcohol is converted into the propargyl ketone by iodide-mediated anodic oxidation, followed by [3 + 2]-cycloaddition of the alkyne unit with azide. The simplicity of the procedure, wide range of substrates, and inexpensive electrode materials are key features of the electrochemical protocol. However, prefunctionalization of the substrate is required [63].

A further electrochemical 1,2,3-triazole synthesis focused on copper catalysis in the context of click chemistry with the generation of copper(I) catalysts or organo-

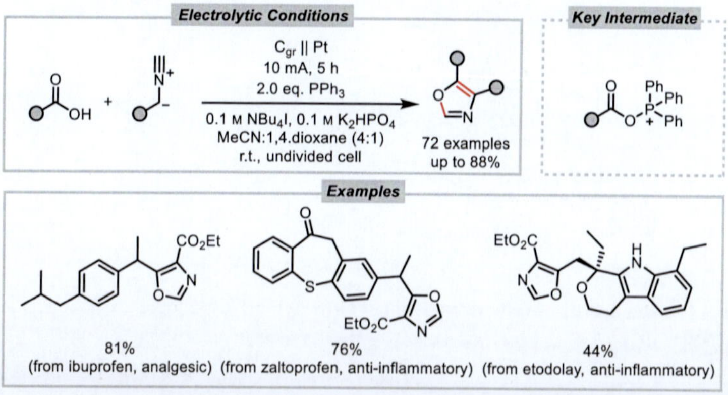

Scheme 8 Electrochemical synthesis of 1,3-oxazoles by [3 + 2]-cycloaddition isocyanides and acyloxyphosphonium ions generated starting from carboxylic acids and phosphine by iodide-mediated anodic oxidation [62]

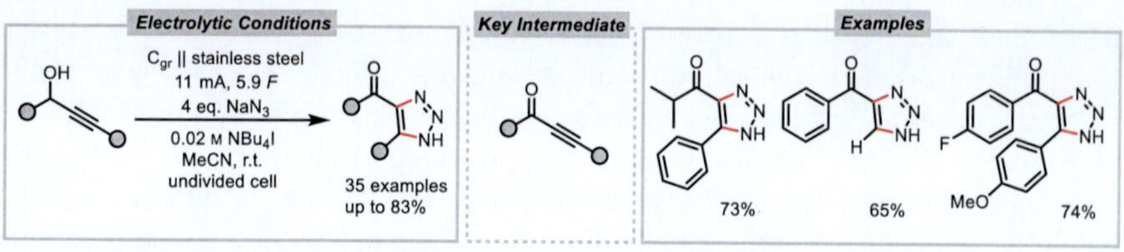

Scheme 9 Electrochemical synthesis of 4,5-disubstituted 1,2,3-triazoles by iodide-mediated anodic oxidation of propargyl alcohols and subsequent [3 + 2]-cycloaddition with sodium azide [63]

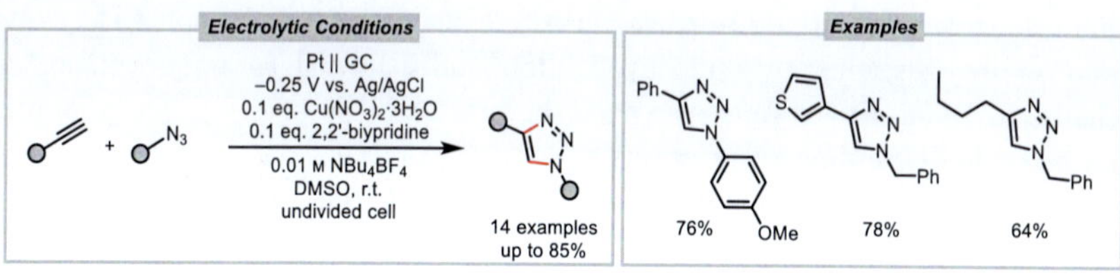

Scheme 10 Potentiostatic in-cell generation of copper(I) catalyst for the synthesis of 1.2.3-triazoles [64]

copper compounds. *Praveen* et al. reported an electrochemical reduction of copper (II) to copper(I) under potentiostatic conditions (Scheme 10). However, 2,2′--bipyridine as ligand and a more sophisticated electrochemical setup is required [64].

The synthesis of copper(I) acetylides is described by *Wilden* et al. by oxidation of acetylene utilizing sacrificial copper anodes under potentiostatic conditions (Scheme 11). This study serves as a promising proof of concept, but its applicability needs to be demonstrated on a broader scale with pharmaceutically relevant compounds. In addition, copper waste is produced in equimolar quantities [65].

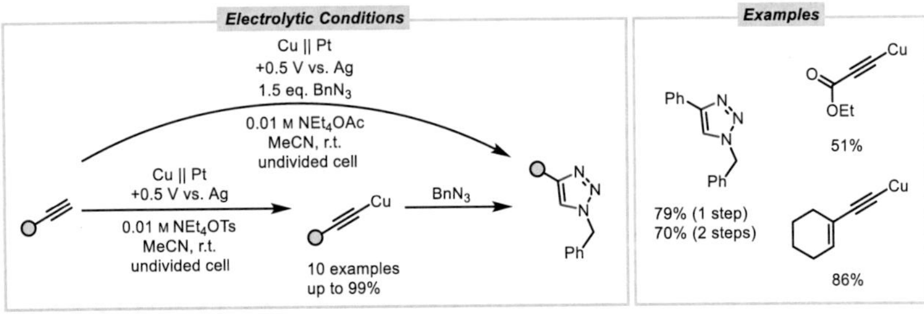

Scheme 11 Potentiostatic synthesis of copper(I) acetylides by anodic oxidation at sacrificial copper anodes [65]

2.2 *Formal [3 + 2]-Cycloaddition in Heterocycle Synthesis*

Electrochemical reactions enable unique reactivity, that is, by the formation of highly reactive radicals. Consequently, the reaction mechanisms are often complex and follow more than one pathway. The following reactions formally represent [3 + 2]-cycloadditions when only considering the formed bonds. However, the essential constraint that they proceed as a pericyclic reaction is by no means fulfilled. Nevertheless, the mechanisms are not always clarified in full detail.

Tang et al. reported an electrochemical synthesis of condensed 1,2,4-triazoles by halide-mediated anodic oxidation of *N*-tosyl hydrazones with quinolines (Scheme 12, top). After iodide-mediated anodic oxidation of the hydrazone, cyclization occurs in two steps and is therefore nonpericyclic. Aromatization to the 1,2,4-triazole is obtained by TsH elimination. The method could be extended to isoquinolines [66]. In addition, *Tang* et al. reported the synthesis of 5-aminotriazoles from diaryl hydrazones with cyanamide under adapted electrolytic conditions by a similar two-step cyclization under formal [3 + 2]-cycloaddition (Scheme 12, bottom). To further expand the practicability of these electrochemical conditions, multicomponent reactions were also examined with a mixture of aldehydes, hydrazines, and cyanamide under the standard conditions [67].

An electrochemical synthesis of 3-aroylindoles from protected anilines and 1,3-diketones was reported by *Sun* et al. (Scheme 13). First, anodic C–C cross-coupling occurred, followed by cyclo-condensation of the aniline with a carbonyl group. The electrochemical method enables the conversion of widely used anilines and 1,3-diaryl-1,3-dicarbonyl compounds under simple electrolytic conditions. A broad scope of highly substituted indoles was demonstrated. However, the electrolytic conditions require HFIP as solvent [68].

Li et al. reported an electrochemical synthesis of 1,2,4-triazolidines from azobenzenes and hexahydro-1,3,5-triazines by formal [3 + 2]-cycloaddition (Scheme 14). First, azobenzene is reduced to *N*, *N′*-diphenylhydrazine at the platinum cathode. Condensation with the triazane or its decomposition products such as imines and formaldehyde yields 1,2,4-triazolidines. The simple and environmentally benign electrochemical protocol was applied on a broad scope [69].

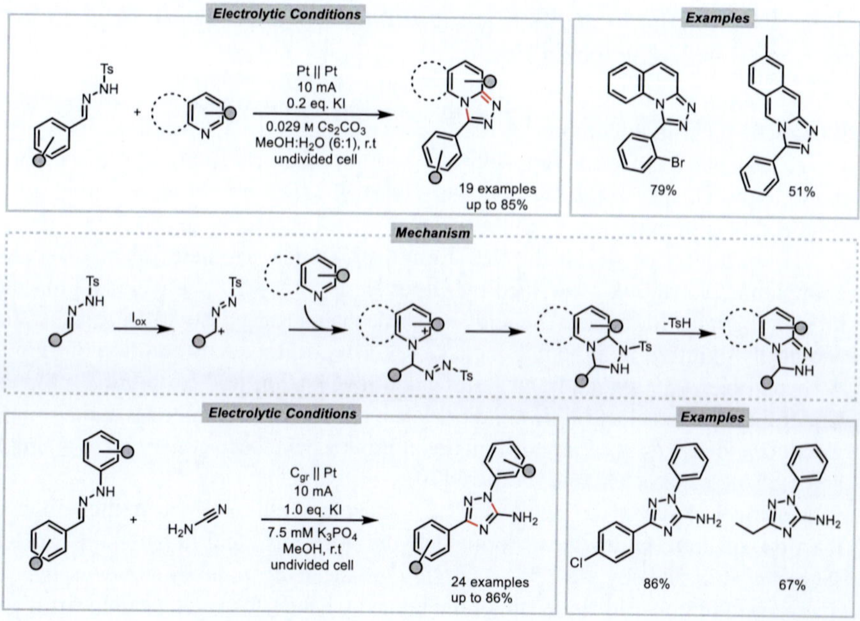

Scheme 12 Electrochemical synthesis of 1,2,4-triazoles by formal [3 + 2]-cycloaddition with quinolines and cyanamide [66, 67]

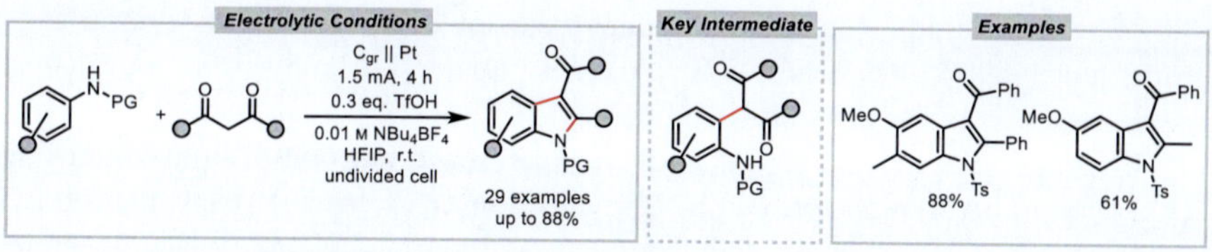

Scheme 13 Electrochemical synthesis of multisubstituted indoles from 1,3-diketones and protected anilines by formal [3 + 2]-cycloaddition [68]

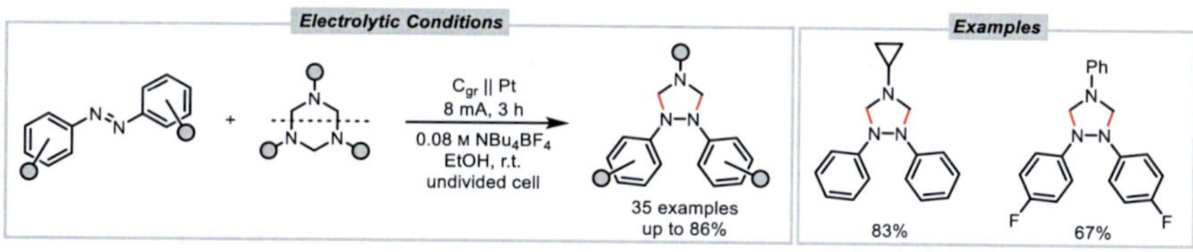

Scheme 14 Electrochemical synthesis of 1,2,4-triazolidine from azobenzenes and 1,3,5-triazanes by formal [3 + 2]-cycloaddition [69]

2.3 Electrochemical Multicomponent Reactions Including a [3 + 2]-Cycloaddition

Electrochemical multicomponent reactions involving a [3 + 2]-cycloaddition as a key step for heterocycle formation are a powerful synthetic technique as they involve multiple in-cell reactant generation steps as well as cascade reactions in a single synthetic operation [17].

The synthesis of 1,2,4-triazoles from hydrazones, aromatic aldehydes, and ammonium acetate was described by *Gu* et al. as a [3 + 1 + 1]-cycloaddition due to in-cell imine formation (Scheme 15). Mechanistically, the hydrazone is first anodically oxidized in an iodide-mediated process, followed by addition of imines. A second benzylic oxidation and subsequent cyclization lead to the final product. Under the reported electrolytic conditions, a wide range of hydrazones and aromatic aldehydes tolerating a variety of functional groups have been converted in a simple and scalable electrochemical protocol [70].

Moreover, *Yuan* et al. reported a 1,2,4-triazole synthesis by a multicomponent reaction of aryl hydrazines, formaldehyde, alcohols, and ammonium acetate (Scheme 16). Thereby, in-cell generation of the hydrazone by condensation of hydrazine with formaldehyde and of the imine by halide-mediated alcohol oxidation and condensation with ammonia is combined. The mechanism of this cyclization was described similarly to previous work from *Gu* et al. by a formal [3 + 2]-cycloaddition initiated by iodide-mediated anodic hydrazone oxidation [71].

Ye et al. reported an electrochemical synthesis of *N*-benzyl tetrazoles involving [3 + 2]-cycloaddition by a multicomponent reaction of alkyl arenes, acetonitrile, and trimethylsilylazide (Scheme 17). Alkenes could be converted into vicinal azidotetrazoles under the same conditions [72].

Masson et al. reported an electrochemical cascade reaction to imidazolines and oxazolidines involving a formal [3 + 2]-cycloaddition (Scheme 18). By direct anodic oxidation of sodium trifluoromethylsulfinate, trifluoromethyl radicals are generated,

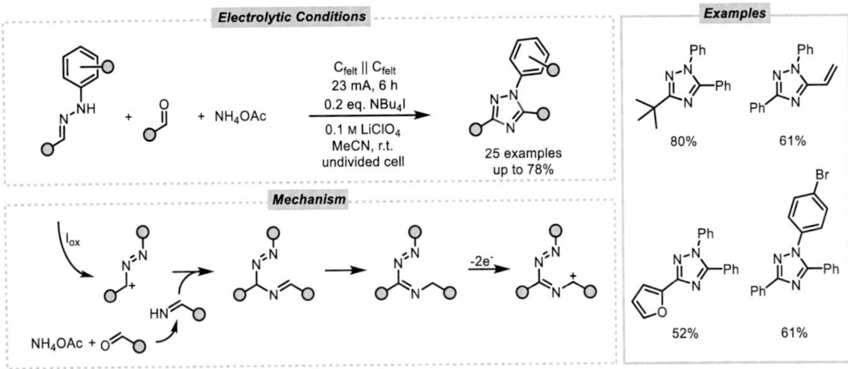

Scheme 15 Electrochemical synthesis of 1,2,4-triazoles by multicomponent reaction involving formal [3 + 2]-cycloaddition [70]

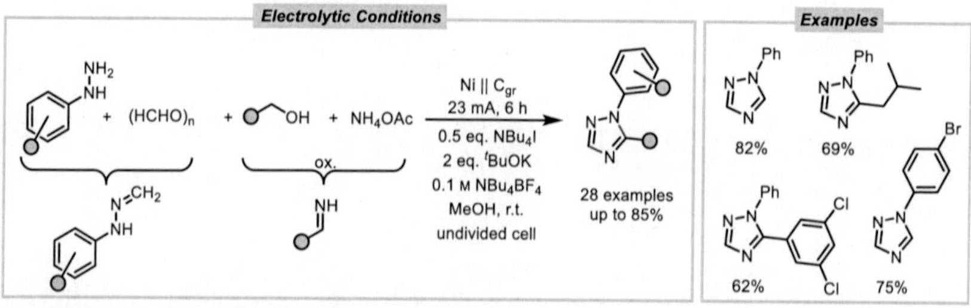

Scheme 16 Electrochemical synthesis of a 1,2,4-triazole synthesis by a multicomponent reagent of hydrazines, formaldehyde, alcohols, and ammonium acetate involving a formal [3 + 2]-cycloaddition [71]

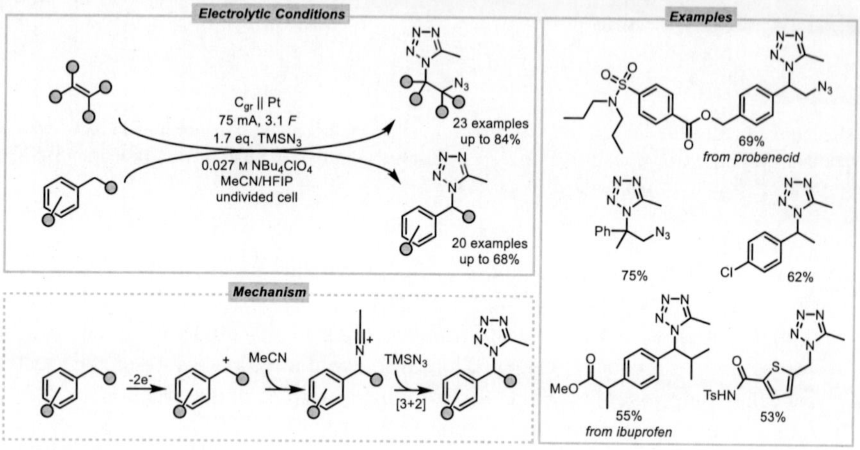

Scheme 17 Electrochemical synthesis of tetrazoles and vicinal azidotetrazoles involving a [3 + 2]-cycloaddition [72]

followed by radical trifluoromethylation of allylamines. A second oxidation of the resulting benzyl radical to the corresponding cation then forms an aziridine, which was identified as a key intermediate obtained by ionic cyclization. Lewis acid-induced ring opening leads to the formal [3 + 2]-cycloaddition with acetonitrile or acetone [73].

In summary, multiple electrochemically induced [3 + 2]-cycloaddition reactions have been reported in recent years, highlighting the practicability and great utility of electro-organic synthesis. However, not all reported electrochemical transformations strictly follow a pericyclic mechanism. Therefore, these reactions are only formally assigned as [3 + 2]-cycloaddition reactions, since the patterns of bonds formed over the course of these reactions are effectively identical to those of a cycloaddition. Electrochemical protocols proved to offer a powerful tool for the formation of 1,3-dipolar compounds as well as dipolarophiles. Furthermore, electrochemical strategies were presented to conveniently activate the reaction partners for the cycloaddition directly or by means of copper catalysis. In addition, primary insights

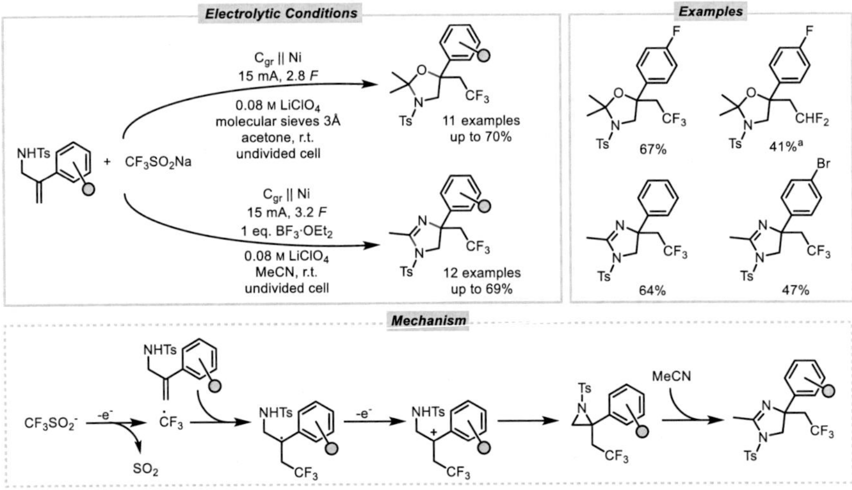

Scheme 18 Electrochemical cascade reaction to imidazolines and oxazolidines induced by CF_3 radical generation involving a formal [3 + 2]-cycloaddition. [a] Sodium difluoromethylsulfinate instead of sodium trifluoromethylsulfinate [73, 74]

into the mechanism and the nature of key intermediates are provided. Nevertheless, improving the knowledge of oxidized reactants is crucial for further optimization of the strategies presented herein, ultimately leading to the implementation of these reactions into technical processes. Moreover, the unique reactivity of the electrochemical transformations could, for example, facilitate the formation of further 1,3-dipolar compounds or the in-cell generation of radicals or highly reactive intermediates, making further applications in heterocycle synthesis very likely.

3 Anodic N–N and N–C Bond Formation in Making Heterocycles

Electrochemistry offers a unique tool for the modern organic chemist to access novel reactivities toward the direct formation of C–N or N–N bonds by anodic oxidation. Various modern approaches to directly obtain heterocycles like indazolin-3-one, dihydroquinoline-2-ones, and many more have been reported in the recent literature, underlining the importance of more sustainable and efficient methods for their synthesis. The biggest challenge of the synthesis of most heterocycles is based on their retrosynthetic considerations. Most of the time, hazardous chemicals, such as hydrazine, or equimolar amounts of oxidants are needed. Electrochemistry offers a simple solution by avoiding these reagents and giving access to a variety of novel motifs. Herein, we survey several electrosynthetic techniques for the synthesis and modification of *N*-heterocycles by C–N or N–N bond formation [75].

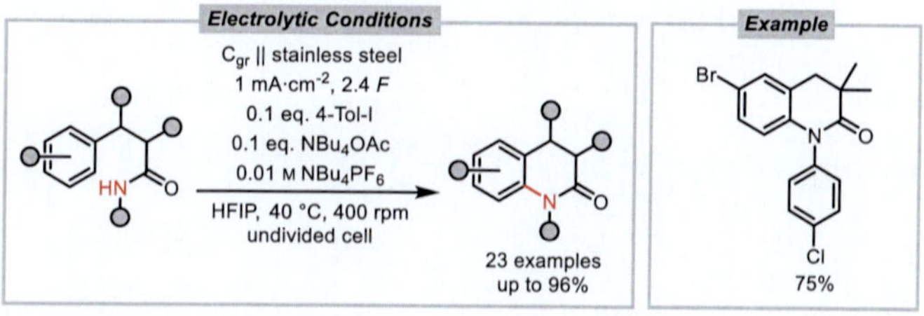

Scheme 19 Electrochemical synthesis of 3,4-dihydroqinoline-2-ones by in situ generated hypervalent iodine [76]

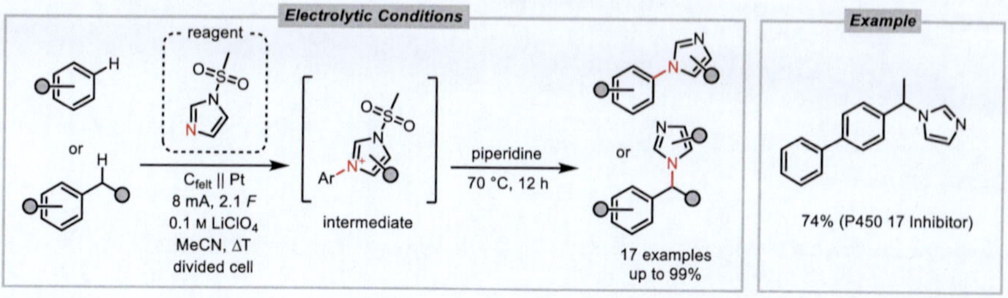

Scheme 20 Electrochemical amination of arenes and benzylic positions by C–H activation [80, 81]

3,4-Dihydroquinolin-2-ones are found in several pharmaceuticals and research chemicals for the treatment of cardiovascular or neurological disorders [77, 78]. The synthesis of dihydroquinazolin-2-ones is commonly achieved by transition metal catalysts based on palladium and copper or by the addition of external oxidizers like PIFA. The electrochemical synthesis of hypervalent iodine species is an established procedure [79]. *Nishiyama* et al. developed the first electrochemical protocol for the synthesis of dihydroquinoline-2-ones by the ex-cell generation of a hypervalent iodine species. This was done in a simple undivided setup using lithium perchlorate as supporting electrolyte. *Waldvogel* et al. established an in-cell protocol that also relies on the electrochemical generation of a hypervalent iodine species starting from 4-iodotoluene (Scheme 19) [76]. The explosive lithium perchlorate was substituted by a quaternary ammonium salt, avoiding harmful supporting electrolytes. By this method, a broad scope of dihydroquinoline-2-ones was obtained. Additionally, the underlying mechanism of this reaction was confirmed by extensive CV studies.

Most organic synthetic methods start from substrates already containing the nitrogen, which is later installed into the corresponding *N*-heterocycle. However, by a simple electrochemical C–H activation, the heterocycle can be constructed from more abundant and less reactive hydrocarbons by in-cell amination and subsequent cyclization. The foundation of this approach was originally established by *Yoshida* et al. by anodic activation of arenes and using a *N*-sulfonyl heterocycle as nitrogen source (Scheme 20) [80–82]. This technique allows for the synthesis of substituted imidazoles in a simple setup by initially oxidizing the corresponding arene and trapping as a cationic heterocycle. After electrolysis, a secondary amine is added,

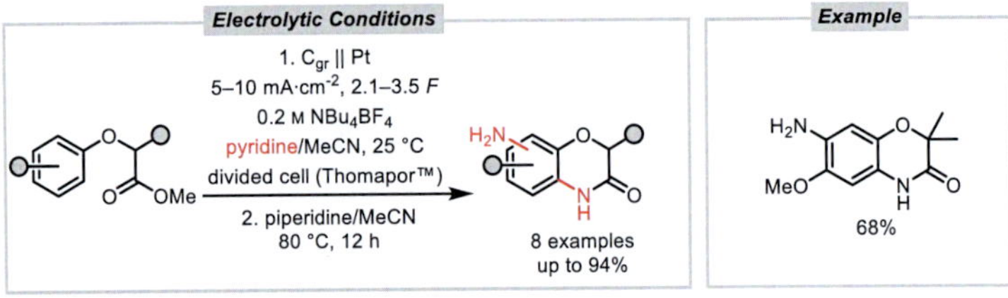

Scheme 21 Electrochemical amination of phenoxy acetates and subsequent cyclo-condensation [83, 84]

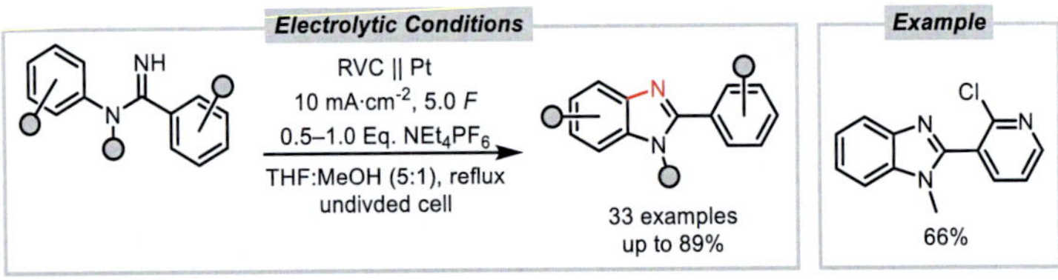

Scheme 22 Electrochemical synthesis of benzimidazoles from aryl amidines [85]

and the mixture is heated to complete the amination by desulfonylation to neutralize the charge on the heterocycle. Furthermore, by variation of the nitrogen source, the modification of arenes resulting in established APIs like fenofibrate or aniracetam was shown. This highlights the importance of this protocol for the manufacturing of pharmaceuticals.

Waldvogel et al. applied a variation of his protocol to phenoxy acetates for the direct synthesis of 1,4-benzoxazin-3-ones with pyridine as readily available nitrogen source in this sequence (Scheme 21) [83, 84]. Initially, a broad variety of 1,4-benzoxazin-3-ones was prepared with diverse functional groups. Following this, the electrolytic protocol was further improved to the twofold amination variant, which gave access to unprotected amino-1,4-benzoxazin-3-ones. This family of compounds might serve as precursors for the development of novel anticancer drugs, as investigated in this study.

Benzimidazoles are a family of compounds with bioactive properties and are used as functional materials. A common synthetic route involves the use of large amounts of transition metals like copper or palladium [86]. *Xu* et al. established a simple synthetic route to various substituted benzimidazoles under mild conditions starting from easily accessible amidines (Scheme 22) [85]. Initially, the amidine is oxidized to a radical, which subsequently forms the corresponding benzimidazole. Thirty-three examples were prepared by these procedures. The mechanism was supported by CV studies.

The synthesis of 6-phenanthridinone is widely used in the manufacturing of dyes and dye precursors. 5*H*-6-Phenanthridinone was first synthesized by *Pictet* and *Ankersmit* by a simple reduction of nitrophenylbenzoic acid with zinc dust and ammonia [88]. Modern methods involve transition metal catalysts or equimolar amounts of redox agents [89, 90]. The first electrochemical synthesis was reported

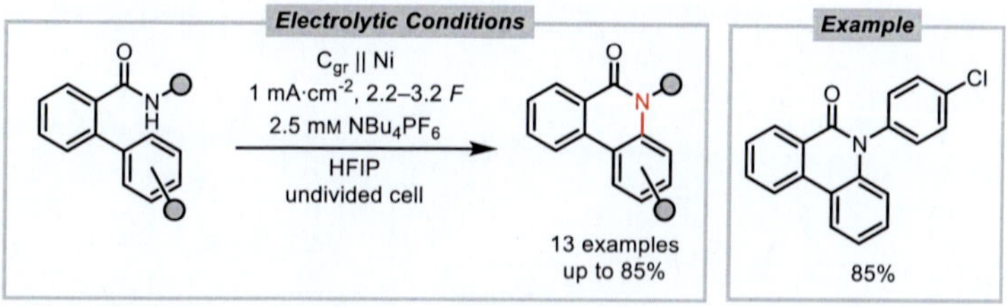

Scheme 23 Electrochemical synthesis of 6-phenanthridinones via activation of biphenyl amides [87]

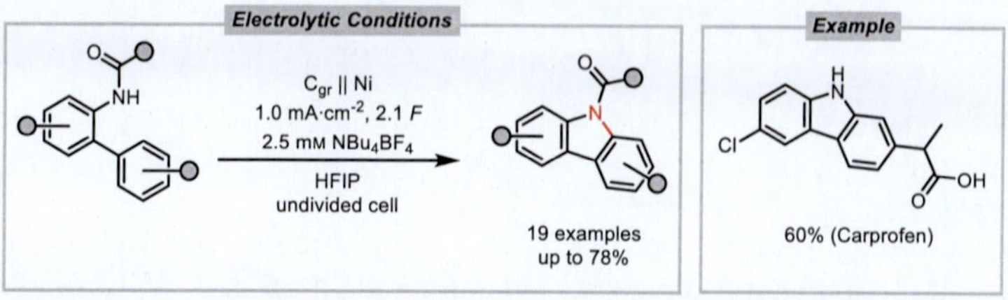

Scheme 24 Electrochemical synthesis of carbazoles from biphenyl anilides [93]

by *Grimshaw* et al. in 1977. An undivided and potential-controlled setup was used with a mercury cathode and a platinum anode. This rather sophisticated setup is not applicable to larger scale and uses noncontemporary and toxic mercury [91]. The first constant-current protocol was established by *Zhang*, *Zeng*, and *Xu* in 2018 using a bromide-mediated system in an undivided setup [92]. However, a large access of bromide was used and large quantities of mediator are not desired. *Waldvogel* et al. established a simple undivided protocol without mediator and a very low supporting electrolyte concentration (Scheme 23) [87]. Thirteen diverse examples were prepared, and the applicability of the method was shown on gram scale. The graphite and nickel used as electrodes are inexpensive, which further underlines the simplicity of the method.

Carbazoles are widely found as a scaffold in various pharmaceuticals, for example, carprofen [94]. Carbazole was first characterized after isolation from high-boiling coal by distillation [95]. Most synthetic approaches involve transition-metal catalysis or equimolar amounts of terminal oxidants [96, 97]. *Nishiyama* et al. established an electrochemical protocol by an ex-cell generation of a hypervalent iodine species that adds to the starting material, forming the desired carbazole. However, lithium perchlorate was used as supporting electrolyte in large quantities [98]. *Francke* et al. developed an in-cell approach with a recyclable mediator-electrolyte system [99]. The mediator served a dual role as iodine source and supporting electrolyte bearing a quaternary ammonium function. *Waldvogel* et al. developed a mediator-free system using minimal amounts of supporting electrolyte (Scheme 24) [93]. They obtained a broad scope of carbazoles in up to very good yields. In addition, the synthesis of carprofen was shown after a simple hydrolysis of the corresponding *N*-acyl carbazole.

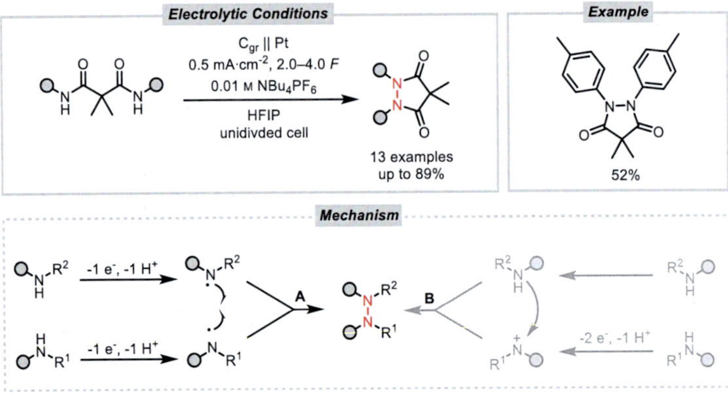

Scheme 25 Electrochemical N–N bond formation by oxidative activation of diamides [100]

Pyrazolidin-3,5-diones were discovered by *E. Fischer* and later described as having anti-inflammatory, analgesic, and antipyretic properties. Most synthetic protocols involve the use of highly toxic hydrazines with the N–N bond already established. *Waldvogel* et al. could achieve the N–N bond formation by a simple electrochemical approach (Scheme 25, top) [100]. The use of hydrazines was avoided by a simple oxidation of amides to amidyl radicals, which formed the N–N bond by combination. Additionally, very low supporting electrolyte concentrations were used, allowing access to several aryl-substituted pyrazolidin-3,5-diones. Even mixed systems with different substituents on the nitrogen atoms were effective.

The possible mechanism of the electrochemical N–N bond formation was extensively studied via CV studies [100]. Since the formation of a N–N bond involves two nitrogen atoms, a diradical (Scheme 25, pathway A) or a cationic pathway (Scheme 25, pathway B) for their connection is possible. Single-electron transfers from both nitrogen atoms form the diradical, which combines to the desired N–N bond. The cationic pathway starts by oxidizing one of the nitrogen atoms to the corresponding cation, which is attacked by a second nucleophilic nitrogen atom to form the N–N bond. The reaction conditions and the intrinsic features of the substituents at the nitrogen atoms strongly determine the reaction pathway of the N–N bond formation.

If thioamides are used in analogous way, no N–N bond formation is observed, but disulfide-containing heterocycles are constructed instead. Thiaheterocycles are found in natural compounds like α-lipoic acid [101]. Their conventional synthesis is accomplished by oxidation of thiols and reduction of sulfonyl halides or by using strong bases [102, 103]. However, conventional routes generate large amounts of waste or use equimolar amounts of hazardous reagents. *Waldvogel* et al. established an easily applicable electrochemical protocol in an undivided cell by adapting the electrolytic conditions of the synthesis of pyrazolidin-3,5-diones (Scheme 26). Methanol as a green solvent was used and a diverse scope of 3,5-diimido-1,2-dithiolanes was prepared in very good yields. Overall, this protocol supplements the electrochemical access to various heterocycles with its unique reactivity.

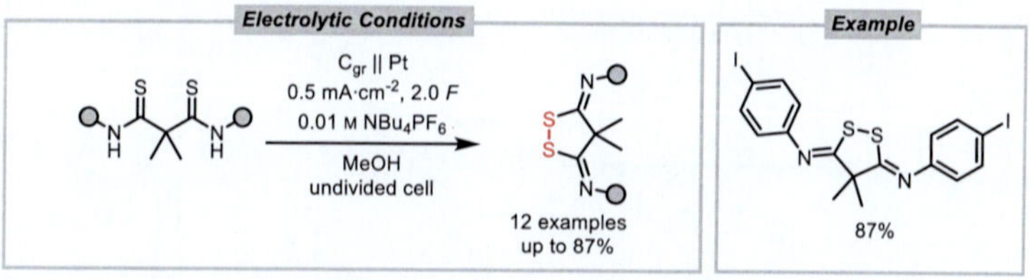

Scheme 26 Electrochemical synthesis of 3,5-diimido-1,2-dithiolanes [101]

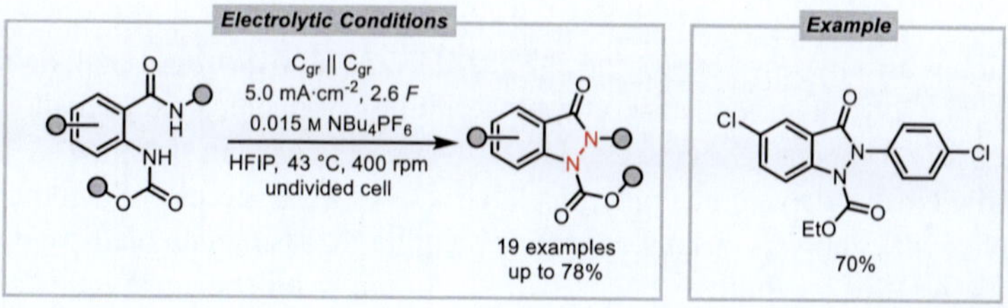

Scheme 27 Simple and scalable protocol for the electrochemical synthesis of novel indazolin-3-ones [104]

Indazolin-3-ones are a group of highly biologically active compounds bearing an endocyclic N–N bond [105, 106]. Their synthesis is commonly achieved by using equimolar amounts of oxidizers like PIFA or harsh reaction conditions involving strong bases [107, 108]. *Kurth* et al. reported a photochemical approach to this motif starting from 2-nitrobenzylalcohols. The first electrochemical access to indazolin-3-ones was obtained by *Trazza* et al. [109]. However, mercury electrodes were used in a complicated potentiostatic reaction setup. Furthermore, potentially carcinogenic nitrosamines were used as substrates [110]. To circumvent the complex setup and the hazardous chemicals, *Waldvogel* et al. developed a modern protocol in a simple galvanostatic setup starting from anthranilic acids (Scheme 27) [104]. Carbon-based electrodes were used to obtain 19 diverse examples. The anodic double oxidation of the substrates led to the formation of the *N*-centered radicals that combine to form the desired indazolin-3-ones. 1,1,1,3,3,3-Hexafluoroisopropanol (HFIP) was used as the solvent due to its stabilizing effect on radical intermediates.

Waldvogel et al. [111] developed a novel protocol for the synthesis of *N,N′*-diarylhydrazines starting from inexpensive and readily available anilines, which are functionalized to their corresponding anilides by a simple reaction with an acid anhydride or acid chloride. Furthermore, carbon electrodes and an acetate buffer were used, highlighting the sustainability of the method. A broad scope was prepared, and subsequent removal of the acyl moieties by simple saponification resulted in the free diphenylhydrazine. This methodology was adapted for the synthesis of phthalazin-1,4-diones.

Phthalazin-1,4-diones are scaffolds having antipyretic or antibacterial properties and are used as important precursors [113]. In general, these motifs are constructed by condensation of phthalic acid and hydrazines, by transition metal catalysis, or by

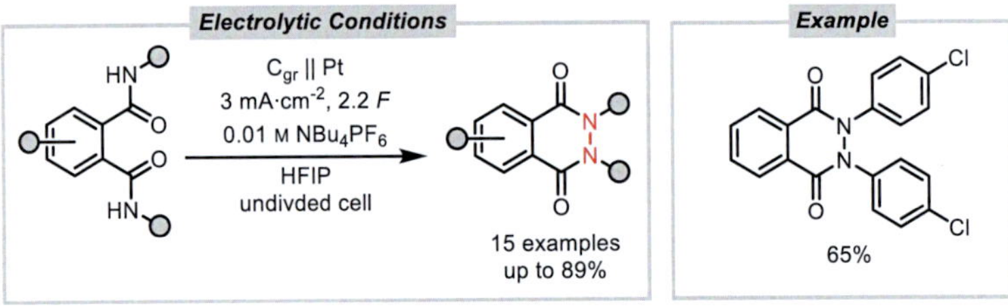

Scheme 28 Electrochemical N-N bond formation boy oxidative activation of phthalamides [112]

oxidation with a hypervalent iodine species [114, 115]. *Waldvogel* et al. developed a simple method under constant current conditions to prepare a broad scope of phthalazin-1,4-diones (Scheme 28) [112]. The products were obtained in moderate to good yields, and even unsymmetrical derivatives were successfully converted. By a simple hydrazinolysis, a substituted azo compound was prepared after electrolysis. This allows this method to be used as common synthetic technique to prepare unsymmetrical azo compounds, which can be later reduced to hydrazines.

The high demand for electrochemical syntheses of various *N*-heterocycles by N–N or C–N bond formation as described herein highlights the significance of this topic in the electrochemical society. Moreover, more modern approaches avoided the use of harsh reaction conditions, equimolar oxidizers, or scarce metal catalysts. Furthermore, the unique reactivity of electrochemical methods allows direct access to several *N*-heterocycles bearing a N–N bond or C–N bond previously not accessible by common routes.

4 Cathodic Conversion of Nitro Moieties to Construct Heterocycles

In the following section, recent examples for heterocycle syntheses via cathodic reduction of nitro groups will be discussed, focusing on galvanostatic methods. Potentiostatic reactions will be addressed only briefly, as well as methods relying on the use of mercury pool electrodes due to the toxicity of mercury. For overviews covering these reactions in more detail, consider the contributions of *Moinet* and *Lund* [116, 117]. Furthermore, the focus will be on the construction of five- and six-membered heterocycles. Reports of the formation of larger cyclic systems remain scarce with limited synthetic usability; hence only electroanalytical studies were undertaken [118]. The electrochemical conversion of functional groups, such as nitro moieties, presents a powerful tool for heterocycle synthesis. In general, the outcome in the electrochemical reduction of nitro groups depends on the total amount of applied charge and the acidity of the reaction medium. Accordingly, nitroso species, hydroxylamines, and even amines are accessible via cathodic reduction of nitro groups (Scheme 29) [18]. Among these, hydroxylamines and amines are prone to nucleophilic ring-closing reactions, whereas nitroso groups generally react

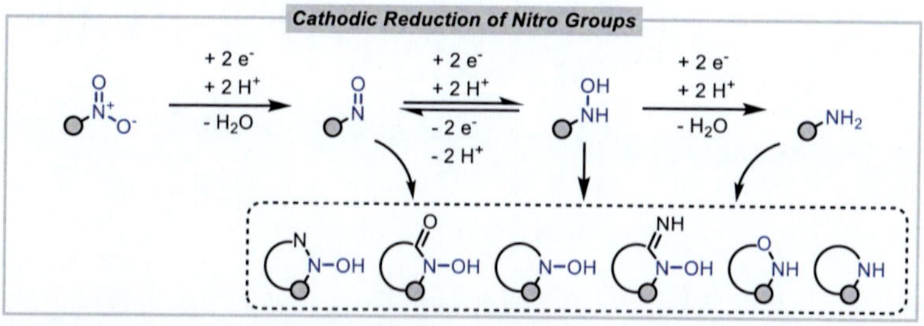

Scheme 29 General mode for the electrochemical reduction of nitroarenes for heterocycle syntheses

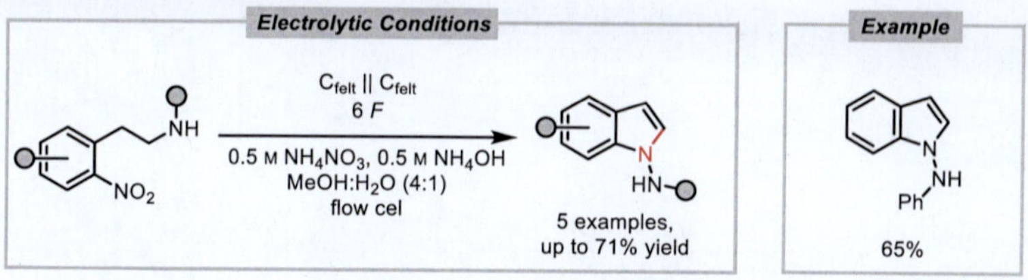

Scheme 30 Electrochemical synthesis of *N*-amino indoles [124]

as *N*-electrophiles. All these reduced species have found applicability in the context of electrochemical heterocycle syntheses [18, 19, 119].

4.1 Construction of Five-Membered Heterocycles

From the manifold of *N*-heterocyclic structures, indoles have a prominent role due to their relevance in medicinal chemistry as well as due to their frequent occurrence in natural products [120, 121]. This important structural motif has been accessed electrochemically from *o*-nitrostyrene derivatives by potentiostatic reduction on a RVC cathode as reported by *Peters* [122, 123]. Additionally, the group of *Moinet* reported the preparation of *N*-aminoindoles through the reduction of 2-(*o*-nitrophenyl)ethylamines with controlled current (Scheme 30) [124]. The proposed mechanism relies on the generation of an aryl nitroso compound, which acts as an electrophile in the nucleophilic ring-closure of the ethylamine group. Thereby, a cinnoline is formed, which is proposed to subsequently rearrange under extrusion of water, yielding the *N*-aminoindole. In total, five examples were prepared in good yields of up to 71%. The reactive nitroso intermediate is accessed stepwise by reducing the nitro group to a hydroxylamine, followed by a subsequent reoxidation. Furthermore, *Moinet* and coworkers made use of a specific flow electrolysis cell, which is divided into three compartments with alternating polarity of the porous carbon electrodes [125].

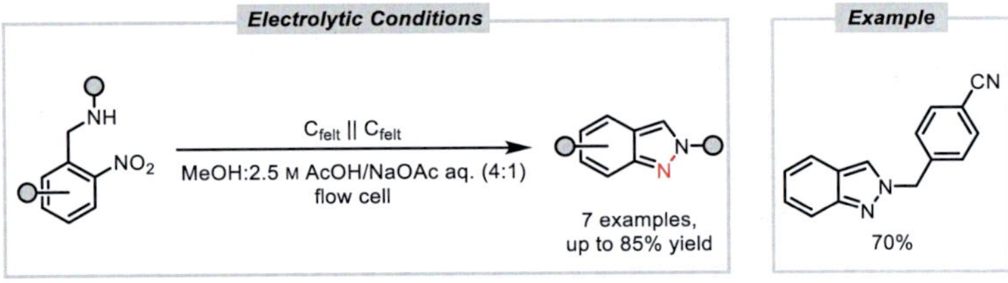

Scheme 31 Electrochemical synthesis of indazoles [126]

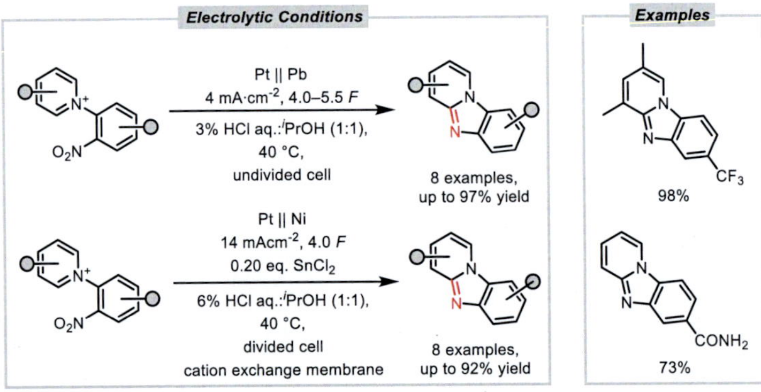

Scheme 32 Electrochemical synthesis of pyrido[1,2-a]benzimidazoles [129, 130]

The same group used the identical cell design for the preparation of indazoles from *o*-nitrobenzylamines (Scheme 31) [126]. As above, the reaction likely proceeds via the stepwise formation of a nitroso species at the porous carbon electrodes. In contrast to their reported indole synthesis, a more acidic electrolyte was used, providing access to seven indazole derivatives with yields up to 85%. Again, potentiostatic methods can also provide access to the indazole motif [109, 127, 128].

Benzimidazoles, which are constitutional isomers of indazoles, were prepared by *Gultyai* et al. starting from *N*-(*o*-nitroaryl)pyridinium salts. In their initial report, a nitroarene is reduced directly at a sacrificial lead cathode in an undivided setup (Scheme 32, top) [129]. The acidic medium is supposed to aid the reduction to a nucleophilic hydroxylamine, which is presumed to react with the neighboring pyridinium salt to close the five-membered heterocyclic core structure of the benzimidazole derivatives. To circumvent the relatively long reaction times resulting from the low current densities, the same group developed a redox-mediated variant (Scheme 32, bottom) [130]. Stannous chloride was found to efficiently catalyze the reduction of the nitro group while tolerating higher current densities, significantly reducing the reaction times. Furthermore, the redox-mediated reaction was conducted without the need for a sacrificial lead electrode, and nickel could be used instead. On the other hand, these improvements are offset by the need for a divided cell, which significantly increases the complexity of the reaction setup. Both variants were used to prepare benzimidazoles in yields greater than 90%. However,

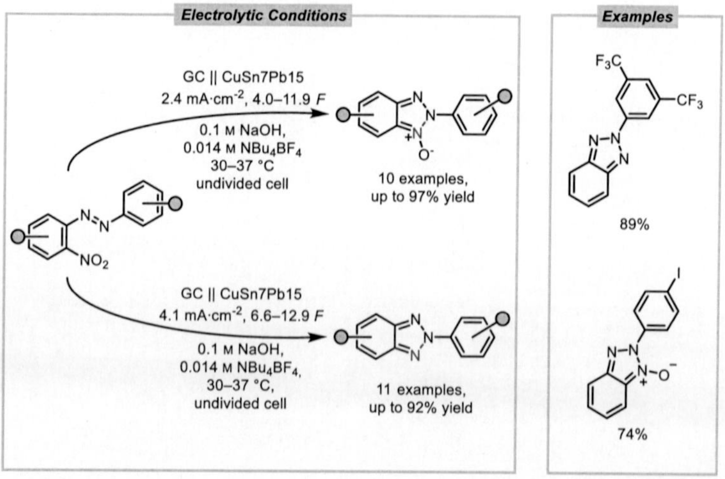

Scheme 33 Electrochemical synthesis of 2-aryl-2*H*-benzo[d][1,2,3]triazoles and their corresponding *N*-oxides [135]

the reported methodologies are restricted to the use of electron-poor nitroarenes, and only eight examples were prepared with either variant. Noteworthily, even the less electron-rich *Zincke* salts obtained by the *Yoshida* amination could not be used [131, 132].

Heterocyclic structures with an even higher count of N atoms are likewise accessible, like benzotriazole derivatives, which were prepared electrochemically as early as 1903 [133]. However, in this initial report, ambiguous experimental details were provided for the potentiostatic synthesis. More recently, another potentiostatic synthesis was reported [134]. *Waldvogel* et al., on the contrary, realized the galvanostatic preparation of benzotriazole derivatives in a simple undivided electrochemical cell from nitro substituted azobenzenes (Scheme 33) [135]. A nitroso species was identified as the reactive intermediate, which is afforded either by direct reduction at the leaded bronze cathode or by a stepwise reduction and subsequent reoxidation. Ring-closing with the neighboring azo group yields the benzotriazole scaffold. Interestingly, the authors showed that the reaction outcome is selectively tunable by the total amount of charge applied and the current density. Applying low current densities results in the formation of triazole *N*-oxides, whereas using higher current densities leads to the reduction of the initially formed *N*-oxide. Both methods provide access to the benzotriazole scaffold in high yields with up to 97%. Remarkably, the authors showed that the supporting electrolyte salt can be omitted while maintaining high yields of the benzotriazoles.

In addition to *N*-heterocyclic scaffolds, *N,O*-heterocyclic 5-membered rings are also available through cathodic reduction of nitro groups. For the benzisoxazole scaffold, there is a multitude of different approaches by following a direct reduction strategy under potentiostatic conditions. In these attempts, various cathode materials came into use, including lead and carbon-based materials [136–138]. Furthermore, a

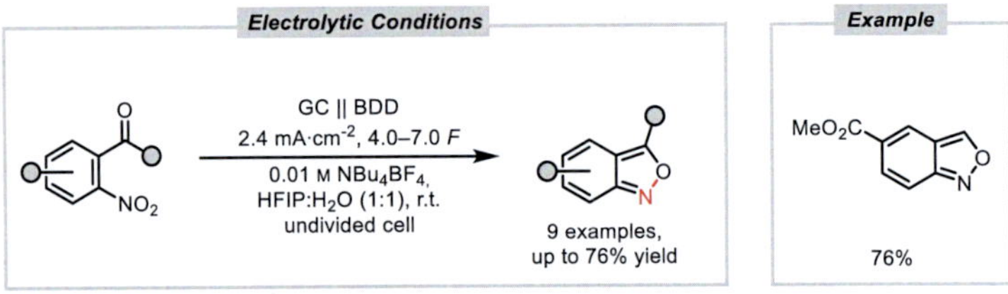

Scheme 34 Electrochemical synthesis of 3-substituted benzo[c]isoxazoles [140]

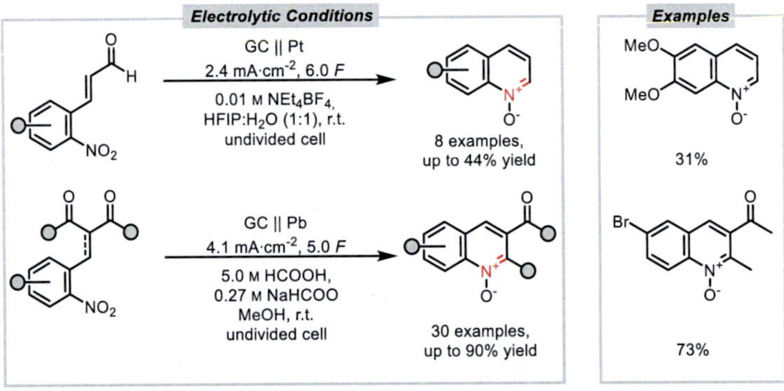

Scheme 35 Electrochemical synthesis of quinolines and their corresponding *N*-oxide [140, 143]

redox-mediated variant based on diselenides is known [139]. A different strategy was pursued in a report by *Waldvogel* et al., wherein galvanostatic conditions were applied to successfully prepare benzisoxazoles from *o*-nitroacetophenone derivatives (Scheme 34) [140]. An undivided cell setup with carbon-based electrodes provided the target structure in yields up to 76% without the need for a mediator.

4.2 Construction of Six-Membered Heterocycles

The construction of *N*-heterocycles is at the forefront of chemical research given their importance in medicinal chemistry [141]. While following electrochemical strategies in this endeavor, not only five-membered heterocyclic structures are accessed but also six-membered scaffolds. A seminal example of such syntheses is the preparation of quinolines from nitroarenes. In an early communication, *Lund* et al. employed a mercury cathode under potentiostatic conditions to prepare quinolines [142]. However, in their polarographic study, only a limited number of different examples were shown to be productive. The *Waldvogel* group, on the contrary, reported the galvanostatic reduction of a nitroarene to yield quinoline *N*-oxides in medium yields of up to 44% (Scheme 35, top) [140]. The conditions employed are analogous to their method for the reductive preparation of

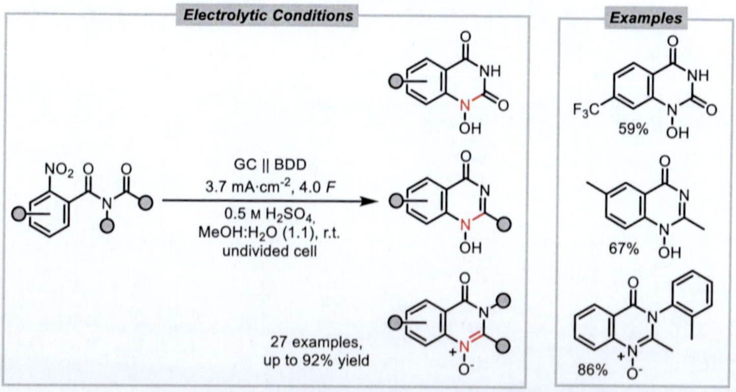

Scheme 36 Electrochemical synthesis of 1*H*-1-hydroxy-quinazolin-4-ones [147]

benzisoxazoles [140]. Supposedly, the nitroarene is reduced at a platinum cathode in an undivided cell to a hydroxylamine derivative, which undergoes condensation with an adjacent aldehyde, resulting in the formation of a quinoline *N*-oxide. Based on these primary investigations, the same group expanded their research efforts [143]. A great range of different quinoline *N*-oxides in yields up to 90% were provided by changing the electrolyte to a buffered acidic medium (Scheme 35, bottom). Furthermore, a lead cathode was chosen to efficiently yield the target structure. It is noteworthy that methods exist to reduce the such prepared quinoline *N*-oxides to the respective quinolines [144].

Moreover, quinazolines were likewise accessible by means of electrochemical cathodic reductions. Whereas early examples relied on potentiostatic reactions and a setup including mercury pool electrodes [145, 146], the *Waldvogel* group more recently reported the synthesis of quinazoline derivatives by using galvanostatic methods in an undivided cell setup (Scheme 36) [147]. The authors employed carbon-based electrode materials for the direct reduction of nitroarenes to hydroxylamines in situ. The acidic reaction medium aids both the reduction and the subsequent ring-closing reaction toward the quinazoline scaffold. Thereby, a variety of heterocyclic compounds was prepared with yields up to 92%.

Other six-membered heterocyclic structures than *N*-based cyclic systems such as *N*,*O*- or *N*,*S*-heterocycles are underrepresented in the electrochemical literature. Most publications dealing with their preparation rely on the potentiostatic cathodic reduction of nitro groups. Another drawback is the frequent use of mercury pool electrodes in such endeavors [148–151]. In a complementary approach, carbon-based electrodes were utilized by *Waldvogel* et al. in their galvanostatic synthesis of benzothiadiazines (Scheme 37) [152]. Nitro-substituted sulfonamides were employed in a divided electrochemical cell setup with a simple glass frit separating the anodic from the cathodic compartment. Either a buffered trifluoroacetic acid solution or a sulfuric acid solution was employed as electrolyte. The latter was used for sterically demanding substrates. Thereby, the nitro group is reduced to a hydroxylamine, which engages in an intramolecular nucleophilic ring-closing reaction with

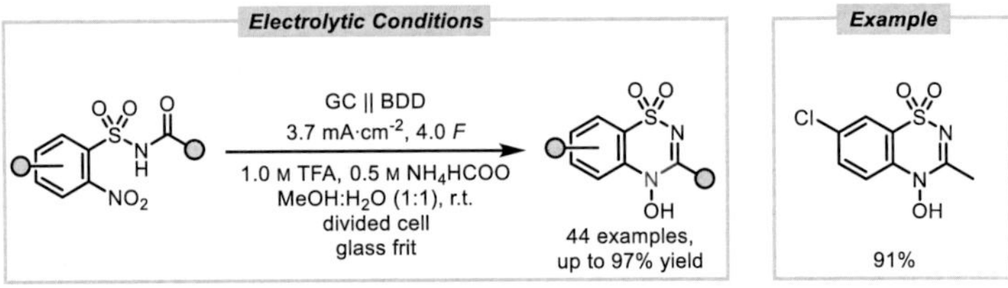

Scheme 37 Electrochemical synthesis of 4-hydroxybenzo[*e*]-1,2,4-thiadiazine-1,1- dioxides [152]

the adjacent sulfonamide. In total, 44 examples with yields up to 97% were accessible through the reported method.

The methods presented herein provide access to a manifold of different five- and six-membered heterocycles by utilizing modern electro-organic methods, namely, the cathodic reduction of nitro groups. In general, this reductive functionalization is a very well-understood electrochemical transformation yielding synthetically worthy and chemically diverse intermediates, such as nitroso compounds, hydroxylamines, and amines. By sophisticated tuning of the reaction conditions, each intermediate can be selectively targeted. It was shown furthermore that every intermediate occurring in this reduction can be harnessed for follow-up reactions, thus rendering the cathodic reduction of nitro groups an efficient method for the synthesis of diverse heterocycles. Moreover, it was demonstrated that methods for cathodic nitro reduction evolved from the use of toxic mercury electrodes in potentiostatic reactions to galvanostatic reactions employing carbon-based electrodes, thereby increasing reproducibility and simplifying the electrochemical setup. In summary, the cathodic reduction of nitro groups has proven to be a valuable method for heterocycle syntheses with high potential for future applications.

5 Perspective

Electro-organic synthesis has experienced a renaissance in this century and has become one of the most valuable tools in modern organic chemistry. Its high synthetic potential is underpinned by its frequent use in the synthesis of high-value products, such as APIs. Furthermore, the reliability and effectiveness of contemporary electro-organic methods set the ground for their successful implementation in the synthesis of natural products. Among the widespread applications, the preparation of structurally diverse and densely functionalized heterocycles is particularly outstanding in these contexts [153–155].

Whereas past electrochemical setups typically lack reproducibility, often due to complex and custom-made equipment, the growing availability of standardized electrochemical setups accelerated the development of new concepts and synthetic methods [33, 38, 156, 157]. Since electro-organic synthesis is a productive field with ongoing research efforts, the toolbox of available methods will continue to evolve. In this regard, the implementation in high-throughput experimentation and the development of automated syntheses in continuous flow systems are two worthwhile research areas, which are promising to impact modern heterocycle synthesis [158–160]. Given the effectivity of electrochemical transformations, further advancing the application of electro-organic methods in industrial processes is beneficial in the pursuit of higher sustainability [161–164].

Acknowledgments The authors acknowledge funding by the DFG (Wa1276/31-1).

References

1. Waldvogel SR, Janza B (2014) Renaissance of electrosynthetic methods for the construction of complex molecules. Angew Chem Int Ed 53:7122–7123. https://doi.org/10.1002/anie.201405082
2. Pollok D, Waldvogel SR (2020) Electro-organic synthesis – a 21st century technique. Chem Sci 11:12386–12400. https://doi.org/10.1039/d0sc01848a
3. Yan M, Kawamata Y, Baran PS (2017) Synthetic organic electrochemical methods since 2000: on the verge of a renaissance. Chem Rev 117:13230–13319. https://doi.org/10.1021/acs.chemrev.7b00397
4. Zhu C, Ang NWJ, Meyer TH et al (2021) Organic electrochemistry: molecular syntheses with potential. ACS Cent Sci 7:415–431. https://doi.org/10.1021/acscentsci.0c01532
5. Shatskiy A, Lundberg H, Kärkäs MD (2019) Organic electrosynthesis: applications in complex molecule synthesis. ChemElectroChem 6:4067–4092. https://doi.org/10.1002/celc.201900435
6. Wiebe A, Gieshoff T, Möhle S et al (2018) Electrifying organic synthesis. Angew Chem Int Ed 57:5594–5619. https://doi.org/10.1002/anie.201711060
7. Gomtsyan A (2012) Heterocycles in drugs and drug discovery. Chem Heterocycl Compd 48: 7–10. https://doi.org/10.1007/s10593-012-0960-z
8. Jampilek J (2019) Heterocycles in medicinal chemistry. Molecules 24:3839. https://doi.org/10.3390/molecules24213839
9. Taylor RD, MacCoss M, Lawson ADG (2014) Rings in drugs. J Med Chem 57:5845–5859. https://doi.org/10.1021/jm4017625
10. Vitaku E, Smith DT, Njardarson JT (2014) Analysis of the structural diversity, substitution patterns, and frequency of nitrogen heterocycles among U.S. FDA approved pharmaceuticals. J Med Chem 57:10257–10274. https://doi.org/10.1021/jm501100b
11. Zhang TY (2017) The evolving landscape of heterocycles in drugs and drug candidates. In: Scriven E, Ramsden C (eds) Heterocyclic chemistry in the 21st century – a tribute to Alan Katritzky, vol 121. Elsevier, pp 1–12
12. Jiang Y, Xu K, Zeng C (2018) Use of electrochemistry in the synthesis of heterocyclic structures. Chem Rev 118:4485–4540. https://doi.org/10.1021/acs.chemrev.7b00271
13. Aslam S, Sbei N, Rani S et al (2023) Heterocyclic electrochemistry: renewable electricity in the construction of heterocycles. ACS Omega 8:6175–6217. https://doi.org/10.1021/acsomega.2c07378
14. Devi S, Jyoti K et al (2022) Electro-organic synthesis: an environmentally benign alternative for heterocycle synthesis. Org Biomol Chem 20:5163–5229. https://doi.org/10.1039/d2ob00572g
15. Liang S, Zeng C-C (2020) Organic electrochemistry: anodic construction of heterocyclic structures. Curr Opin Electrochem 24:31–43. https://doi.org/10.1016/j.coelec.2020.06.005

16. Sbei N, Listratova AV, Titov AA et al (2019) Recent advances in electrochemistry for the synthesis of N-heterocycles. Synthesis 51:2455–2473. https://doi.org/10.1055/s-0037-1611797
17. Imeni S, Makarem A, Javahershenas R (2023) Recent advances in multicomponent electro-organic (electrochemical) synthesis of heterocycles. Asian J Org Chem 12:e202300303. https://doi.org/10.1002/ajoc.202300303
18. Wirtanen T, Rodrigo E, Waldvogel SR (2020) Recent advances in the electrochemical reduction of substrates involving N–O bonds. Adv Synth Catal 362:2088–2101. https://doi.org/10.1002/adsc.202000349
19. Francke R (2014) Recent advances in the electrochemical construction of heterocycles. Beilstein J Org Chem 10:2858–2873. https://doi.org/10.3762/bjoc.10.303
20. Francke R (2022) Concepts for sustainable organic electrosynthesis. Curr Opin Electrochem 36:101111. https://doi.org/10.1016/j.coelec.2022.101111
21. Francke R, Little RD (2014) Redox catalysis in organic electrosynthesis: basic principles and recent developments. Chem Soc Rev 43:2492–2521. https://doi.org/10.1039/c3cs60464k
22. Möhle S, Zirbes M, Rodrigo E et al (2018) Modern electrochemical aspects for the synthesis of value-added organic products. Angew Chem Int Ed 57:6018–6041. https://doi.org/10.1002/anie.201712732
23. Xu H-C, Moeller KD (2021) Electrochemistry in synthetic organic chemistry. J Org Chem 86: 15845–15846. https://doi.org/10.1021/acs.joc.1c02388
24. Lips S, Waldvogel SR (2019) Use of boron-doped diamond electrodes in electro-organic synthesis. ChemElectroChem 6:1649–1660. https://doi.org/10.1002/celc.201801620
25. Einaga Y (2022) Boron-doped diamond electrodes: fundamentals for electrochemical applications. Acc Chem Res 55:3605–3615. https://doi.org/10.1021/acs.accounts.2c00597
26. Zirbes M, Quadri LL, Breiner M et al (2020) High-temperature electrolysis of Kraft Lignin for selective vanillin formation. ACS Sustain Chem Eng 8:7300–7307. https://doi.org/10.1021/acssuschemeng.0c00162
27. Zirbes M, Schmitt D, Beiser N et al (2019) Anodic degradation of lignin at active transition metal-based alloys and performance-enhanced anodes. ChemElectroChem 6:155–161. https://doi.org/10.1002/celc.201801218
28. Gombos LG, Waldvogel SR (2022) Electrochemical bromofunctionalization of alkenes and alkynes – to sustainability and beyond. Sustain Chem 3:430–454. https://doi.org/10.3390/suschem3040027
29. Gombos LG, Nikl J, Waldvogel SR (2024) Dual roles of supporting electrolytes in organic electrosynthesis. ChemElectroChem:e202300730. https://doi.org/10.1002/celc.202300730
30. Liu K, Song C, Lei A (2018) Recent advances in iodine mediated electrochemical oxidative cross-coupling. Org Biomol Chem 16:2375–2387. https://doi.org/10.1039/C8OB00063H
31. Wirth T (2021) Iodine(III) mediators in electrochemical batch and flow reactions. Curr Opin Electrochem 28:100701. https://doi.org/10.1016/j.coelec.2021.100701
32. Francke R (2021) Recent progress in the electrochemistry of hypervalent iodine compounds. Curr Opin Electrochem 28:100719. https://doi.org/10.1016/j.coelec.2021.100719
33. Beil SB, Pollok D, Waldvogel SR (2021) Reproducibility in electroorganic synthesis-myths and misunderstandings. Angew Chem Int Ed 60:14750–14759. https://doi.org/10.1002/anie.202014544
34. Schulz L, Waldvogel S (2019) Solvent control in electro-organic synthesis. Synlett 30:275–286. https://doi.org/10.1055/s-0037-1610303
35. Hilt G (2020) Basic strategies and types of applications in organic electrochemistry. ChemElectroChem 7:395–405. https://doi.org/10.1002/celc.201901799
36. Moeller KD (2018) Using physical organic chemistry to shape the course of electrochemical reactions. Chem Rev 118:4817–4833. https://doi.org/10.1021/acs.chemrev.7b00656
37. Horn EJ, Rosen BR, Baran PS (2016) Synthetic organic electrochemistry: an enabling and innately sustainable method. ACS Cent Sci 2:302–308. https://doi.org/10.1021/acscentsci.6b00091

38. Leech MC, Lam K (2022) A practical guide to electrosynthesis. Nat Rev Chem 6:275–286. https://doi.org/10.1038/s41570-022-00372-y
39. Heard DM, Lennox AJJ (2020) Electrode materials in modern organic electrochemistry. Angew Chem Int Ed 59:18866–18884. https://doi.org/10.1002/anie.202005745
40. Wirtanen T, Prenzel T, Tessonnier J-P et al (2021) Cathodic corrosion of metal electrodes-how to prevent it in electroorganic synthesis. Chem Rev 121:10241–10270. https://doi.org/10.1021/acs.chemrev.1c00148
41. Kulisch J, Nieger M, Stecker F et al (2011) Efficient and stereodivergent electrochemical synthesis of optically pure menthylamines. Angew Chem Int Ed 50:5564–5567. https://doi.org/10.1002/anie.201101330
42. Klein M, Waldvogel SR (2022) Counter electrode reactions-important stumbling blocks on the way to a working electro-organic synthesis. Angew Chem Int Ed 61:e202204140. https://doi.org/10.1002/anie.202204140
43. Huisgen R (1961) Proceedings of the chemical society. Proc Chem Soc:357–396. https://doi.org/10.1039/PS9610000357
44. Breugst M, Reissig H-U (2020) The Huisgen reaction: milestones of the 1,3-dipolar cycloaddition. Angew Chem Int Ed 59:12293–12307. https://doi.org/10.1002/anie.202003115
45. Shono T, Matsumura Y, Tsubata K et al (1989) Electroorganic chemistry. 116. Electrochemical transformation of aldoximes to nitriles using halogen ions as mediators: intermediary formation of nitrile oxides. J Org Chem 54:2249–2251. https://doi.org/10.1021/jo00270a044
46. Holman SDL, Wills AG, Fazakerley NJ et al (2022) Electrochemical synthesis of isoxazolines: method and mechanism. Chem A Eur J 28:e202103728. https://doi.org/10.1002/chem.202103728
47. Hartmer MF, Waldvogel SR (2015) Electroorganic synthesis of nitriles via a halogen-free domino oxidation-reduction sequence. Chem Commun 51:16346–16348. https://doi.org/10.1039/C5CC06437F
48. Gütz C, Stenglein A, Waldvogel SR (2017) Highly modular flow cell for electroorganic synthesis. Org Process Res Dev 21:771–778. https://doi.org/10.1021/acs.oprd.7b00123
49. Hofmann S, Winter J, Prenzel T et al (2023) Electrochemical synthesis of isoxazoles and isoxazolines via anodic oxidation of oximes. ChemElectroChem 10:e202300434. https://doi.org/10.1002/celc.202300434
50. Luginina J, Linden M, Bazulis M et al (2021) Electrosynthesis of stable Betulin-derived nitrile oxides and their application in synthesis of cytostatic Lupane-type triterpenoid-Isoxazole conjugates. Eur J Org Chem 2021:2557–2577. https://doi.org/10.1002/ejoc.202100293
51. Ji S, Zhao L, Miao B et al (2023) Electrochemical activation of nitromethane to construct Isoxazoline Aldoximes. Angew Chem Int Ed 62:e202304434. https://doi.org/10.1002/anie.202304434
52. Linden M, Hofmann S, Herman A et al (2023) Electrochemical synthesis of pyrazolines and pyrazoles via 3+2 dipolar cycloaddition. Angew Chem Int Ed 62:e202214820. https://doi.org/10.1002/anie.202214820
53. Linden M, Hofmann S, Weber FN et al (2023) From screening to the hectogram scale: sustainable electrochemical synthesis of mefenpyr-diethyl. Green Chem 25:6623–6628. https://doi.org/10.1039/D3GC02118A
54. Hofmann S, Linden M, Neuner J et al (2023) Electrochemically enabled oxidative aromatization of pyrazolines. Org Biomol Chem 21:4694–4701. https://doi.org/10.1039/D3OB00671A
55. Guo H, Liu Y, Wan J-P (2023) Electrochemical cascade pyrazole annulation and C-H halogenation for the synthesis of 4-halopyrazoles. Green Synth Catal. https://doi.org/10.1016/j.gresc.2023.10.004
56. Lyalin BV, Petrosyan VA (2014) New approach to electrochemical iodination of arenes exemplified by the synthesis of 4-iodopyrazoles of different structures. Russ Chem Bull 63:360–367. https://doi.org/10.1007/s11172-014-0438-5
57. Lyalin BV, Petrosyan VA (2012) Effect of “anomalous” electrochemical halogenation of pyrazoles and its reasons. Russ Chem Bull 61:209–210. https://doi.org/10.1007/s11172-012-0030-9

58. Lyalin BV, Petrosyan VA, Ugrak BI (2010) Electrosynthesis of 4-iodopyrazole and its derivatives. Russ Chem Bull 59:1549–1555. https://doi.org/10.1007/s11172-010-0277-y
59. Wu Y, Chen J-Y, Ning J et al (2021) Electrochemical multicomponent synthesis of 4-selanylpyrazoles under catalyst- and chemical-oxidant-free conditions. Green Chem 23: 3950–3954. https://doi.org/10.1039/d1gc00562f
60. Zhang Y, Xu S, Zhu Y et al (2023) One-pot synthesis of 4-thiocyanato-1 H-pyrazoles through electrochemical multicomponent thiocyanation under metal- and oxidant-free conditions. Eur J Org Chem 26:e202201278. https://doi.org/10.1002/ejoc.202201278
61. Kumar R, Banerjee P (2021) Electrochemical generation of a nonstabilized Azomethine Ylide: access to substituted N-heterocycles. J Org Chem 86:16104–16113. https://doi.org/10.1021/acs.joc.1c02069
62. Zhang X, Yuan Q, Zhang H et al (2023) Electrochemical synthesis of oxazoles via a phosphine-mediated deoxygenative [3 + 2] cycloaddition of carboxylic acids. Green Chem 25:1435–1441. https://doi.org/10.1039/D2GC04559A
63. Bandyopadhyay M, Bhadra S, Pathak S et al (2023) An atom-economic method for 1,2,3-Triazole derivatives via oxidative 3 + 2 cycloaddition harnessing the power of electrochemical oxidation and click chemistry. J Org Chem 88:15772–15782. https://doi.org/10.1021/acs.joc.3c01836
64. Krishnan M, Kathiresan M, Praveen C (2023) Electrochemically generated copper(I)-catalyzed click chemistry: Triazole synthesis and insights into their Photophysical properties. Eur J Org Chem 26:e202201405. https://doi.org/10.1002/ejoc.202201405
65. Seavill PW, Holt KB, Wilden JD (2019) Electrochemical synthesis of copper(i) acetylides via simultaneous copper ion and catalytic base electrogeneration for use in click chemistry. RSC Adv 9:29300–29304. https://doi.org/10.1039/C9RA06782E
66. Zhang W, Jiang R, Mu Y et al (2023) Electrochemical oxidation intermolecular [3 + 2] cycloaddition of N-tosylhydrazones and quinolone derivatives to access fused 1,2,4-triazole compounds. Tetrahedron Lett 114:154256. https://doi.org/10.1016/j.tetlet.2022.154256
67. Yang M, Jiang R, Mu Y et al (2023) Electrochemical cycloaddition of hydrazones with cyanamide for the synthesis of substituted 5-amine-1,2,4-triazoles. Chem Commun 59: 2303–2306. https://doi.org/10.1039/D2CC06277A
68. Chang X, Chen X, Lu S et al (2022) Electrochemical [3+2] cycloaddition of anilines and 1,3-dicarbonyl compounds: construction of multisubstituted indoles. Adv Synth Catal 364: 2865–2871. https://doi.org/10.1002/adsc.202200488
69. Li C, Zhong Q, Tang S et al (2022) Electrochemical formal [3 + 2] cycloaddition of azobenzenes with hexahydro-1,3,5-triazines. Org Chem Front 9:3769–3774. https://doi.org/10.1039/D2QO00530A
70. Zhao Z, He Y, Li M et al (2021) An electrochemical multicomponent [3 + 1 + 1] annulations to synthesize polysubstituted 1,2,4-triazoles. Tetrahedron 87:132111. https://doi.org/10.1016/j.tet.2021.132111
71. Yang N, Yuan G (2018) A multicomponent electrosynthesis of 1,5-disubstituted and 1-aryl 1,2,4-triazoles. J Org Chem 83:11963–11969. https://doi.org/10.1021/acs.joc.8b01808
72. Yu Y, Zhu X-B, Yuan Y et al (2022) An electrochemical multicomponent reaction toward C-H tetrazolation of alkyl arenes and vicinal azidotetrazolation of alkenes. Chem Sci 13:13851–13856. https://doi.org/10.1039/d2sc05423j
73. Claraz A, Djian A, Masson G (2021) Electrochemical tandem trifluoromethylation of allylamines/formal (3 + 2)-cycloaddition for the rapid access to CF 3-containing imidazolines and oxazolidines. Org Chem Front 8:288–296. https://doi.org/10.1039/d0qo01307b
74. Kisukuri CM, Fernandes VA, Delgado JAC et al (2021) Electrochemical installation of CFH2 -, CF2 H-, CF3 -, and perfluoroalkyl groups into small organic molecules. Chem Rec 21:2502–2525. https://doi.org/10.1002/tcr.202100065
75. Gieshoff T, Kehl A, Schollmeyer D et al (2017) Electrochemical synthesis of benzoxazoles from anilides – a new approach to employ amidyl radical intermediates. Chem Commun 53: 2974–2977. https://doi.org/10.1039/C7CC00927E

76. Bieniek JC, Mashtakov B, Schollmeyer D et al (2024) Dehydrogenative electrochemical synthesis of N-aryl-3,4-dihydroquinolin-2-ones by iodine(III)-mediated coupling reaction. Chem A Eur J 30:e202303388. https://doi.org/10.1002/chem.202303388
77. Clark PGK, Vieira LCC, Tallant C et al (2015) LP99: discovery and synthesis of the first selective BRD7/9 Bromodomain inhibitor. Angew Chem Int Ed 54:6217–6221. https://doi.org/10.1002/anie.201501394
78. Sander GE, Giles TD (1986) Carteolol. Cardiovasc Drug Rev 4:19–39. https://doi.org/10.1111/j.1527-3466.1986.tb00486.x
79. Francke R, Broese T, Roesel AF (2009) Electrochemistry of hypervalent iodine compounds. In: Patai S, Rappoport Z (eds) Patai's chemistry of functional groups. Wiley, Chichester, pp 1–22
80. Morofuji T, Shimizu A, Yoshida J-I (2014) Direct C-N coupling of imidazoles with aromatic and benzylic compounds via electrooxidative C-H functionalization. J Am Chem Soc 136:4496–4499. https://doi.org/10.1021/ja501093m
81. Morofuji T, Shimizu A, Yoshida J-I (2015) Heterocyclization approach for electrooxidative coupling of functional primary alkylamines with aromatics. J Am Chem Soc 137:9816–9819. https://doi.org/10.1021/jacs.5b06526
82. Waldvogel SR, Möhle S (2015) Versatile electrochemical C-H amination via Zincke intermediates. Angew Chem Int Ed 54:6398–6399. https://doi.org/10.1002/anie.201502638
83. Wesenberg LJ, Diehl E, Zähringer TJB et al (2020) Metal-free twofold electrochemical C-H amination of activated arenes: application to medicinally relevant precursor synthesis. Chem A Eur J 26:17574–17580. https://doi.org/10.1002/chem.202003852
84. Wesenberg LJ, Herold S, Shimizu A et al (2017) New approach to 1,4-Benzoxazin-3-ones by electrochemical C-H amination. Chem A Eur J 23:12096–12099. https://doi.org/10.1002/chem.201701979
85. Zhao H-B, Zhuang J-L, Xu H-C (2021) Electrochemical synthesis of benzimidazoles via dehydrogenative cyclization of amidines. ChemSusChem 14:1692–1695. https://doi.org/10.1002/cssc.202100254
86. Brasche G, Buchwald SL (2008) C-H functionalization/C-N bond formation: copper-catalyzed synthesis of benzimidazoles from amidines. Angew Chem Int Ed 47:1932–1934. https://doi.org/10.1002/anie.200705420
87. Kehl A, Breising VM, Schollmeyer D et al (2018) Electrochemical synthesis of 5-Aryl-phenanthridin-6-one by dehydrogenative N,C bond formation. Chem A Eur J 24:17230–17233. https://doi.org/10.1002/chem.201804638
88. Pictet A, Ankersmit HJ (1891) Ueber das Phenanthridin. Justus Liebigs Ann Chem 266:138–153. https://doi.org/10.1002/jlac.18912660107
89. Yuan M, Chen L, Wang J et al (2015) Transition-metal-free synthesis of phenanthridinones from biaryl-2-oxamic acid under radical conditions. Org Lett 17:346–349. https://doi.org/10.1021/ol503459s
90. Geng X, He H, Shatskiy A et al (2023) Construction of phenanthridinone skeletons through palladium-catalyzed annulation. J Org Chem 88:12738–12743. https://doi.org/10.1021/acs.joc.3c01429
91. Grimshaw J, Mannus D (1977) Electrochemical reactions. Part 20. Intramolecular cyclisation during the reduction of 2-chloro-NN-diphenylbenzamides. J Chem Soc Perkin Trans 1:2456. https://doi.org/10.1039/P19770002456
92. Zhang S, Li L, Xue M et al (2018) Electrochemical formation of N-Acyloxy Amidyl radicals and their application: regioselective intramolecular amination of sp2 and sp3 C-H bonds. Org Lett 20:3443–3446. https://doi.org/10.1021/acs.orglett.8b00981
93. Kehl A, Schupp N, Breising VM et al (2020) Electrochemical synthesis of carbazoles by dehydrogenative coupling reaction. Chem A Eur J 26:15847–15851. https://doi.org/10.1002/chem.202003430
94. Sellamuthu S, Gutti G, Kumar D et al (2018) Carbazole: a potent scaffold for antitubercular drugs. Mini-Rev Org Chem 15:498–507. https://doi.org/10.2174/1570193X15666180220141342

95. Graebe C, Glaser C (1872) Ueber carbazol. Justus Liebigs Ann Chem 163:343–360. https://doi.org/10.1002/jlac.18721630305
96. Cho SH, Yoon J, Chang S (2011) Intramolecular oxidative C-N bond formation for the synthesis of carbazoles: comparison of reactivity between the copper-catalyzed and metal-free conditions. J Am Chem Soc 133:5996–6005. https://doi.org/10.1021/ja111652v
97. Tsang WCP, Zheng N, Buchwald SL (2005) Combined C-H functionalization/C-N bond formation route to carbazoles. J Am Chem Soc 127:14560–14561. https://doi.org/10.1021/ja055353i
98. Kajiyama D, Inoue K, Ishikawa Y et al (2010) A synthetic approach to carbazoles using electrochemically generated hypervalent iodine oxidant. Tetrahedron 66:9779–9784. https://doi.org/10.1016/j.tet.2010.11.015
99. Broese T, Francke R (2016) Electrosynthesis using a recyclable mediator-electrolyte system based on ionically tagged phenyl iodide and 1,1,1,3,3,3-hexafluoroisopropanol. Org Lett 18:5896–5899. https://doi.org/10.1021/acs.orglett.6b02979
100. Gieshoff T, Kehl A, Schollmeyer D et al (2017) Insights into the mechanism of anodic N-N bond formation by dehydrogenative coupling. J Am Chem Soc 139:12317–12324. https://doi.org/10.1021/jacs.7b07488
101. Breising VM, Gieshoff T, Kehl A et al (2018) Electrochemical formation of 3,5-diimido-1,2-dithiolanes by dehydrogenative coupling. Org Lett 20:6785–6788. https://doi.org/10.1021/acs.orglett.8b02904
102. Ni C, Hu M, Hu J (2015) Good partnership between sulfur and fluorine: sulfur-based fluorination and fluoroalkylation reagents for organic synthesis. Chem Rev 115:765–825. https://doi.org/10.1021/cr5002386
103. Wang M, Jiang X (2018) Sulfur-sulfur bond construction. Top Curr Chem 376:14. https://doi.org/10.1007/s41061-018-0192-5
104. Bieniek JC, Grünewald M, Winter J et al (2022) Electrochemical synthesis of N,N'-disubstituted indazolin-3-ones via an intramolecular anodic dehydrogenative N-N coupling reaction. Chem Sci 13:8180–8186. https://doi.org/10.1039/D2SC01827F
105. Arán VJ, Flores M, Muñoz P et al (1996) Analogues of cytostatic, fused indazolinones: synthesis, conformational analysis and cytostatic activity against HeLa cells of some 1-substituted indazolols, 2-substituted indazolinones, and related compounds. Liebigs Ann 1996:683–691. https://doi.org/10.1002/jlac.199619960507
106. Foster SJ, Bruneau P, Walker ER et al (1990) 2-substituted indazolinones: orally active and selective 5-lipoxygenase inhibitors with anti-inflammatory activity. Br J Pharmacol 99:113–118. https://doi.org/10.1111/j.1476-5381.1990.tb14663.x
107. Barton DHR, Lukacs G, Wagle D (1982) Reactions of hydrazides and hydrazones with n-butyl-lithium. J Chem Soc Chem Commun:450–452. https://doi.org/10.1039/C39820000450
108. Correa A, Tellitu I, Domínguez E et al (2006) An advantageous synthesis of new indazolone and pyrazolone derivatives. Tetrahedron 62:11100–11105. https://doi.org/10.1016/j.tet.2006.09.031
109. Marrosu G, Petrucci R, Trazza A (1995) Electroreduction of N-benzyl-N-nitrosoanthranilic acid to 1-benzyl-1,2-dihydro-3H-indazol-3-one. Electrochim Acta 40:923–925. https://doi.org/10.1016/0013-4686(94)E0179-4
110. Li K, Ricker K, Tsai FC et al (2021) Estimated cancer risks associated with nitrosamine contamination in commonly used medications. Int J Environ Res Public Health 18:9465. https://doi.org/10.3390/ijerph18189465
111. Breising VM, Kayser JM, Kehl A et al (2020) Electrochemical formation of N,N'-diarylhydrazines by dehydrogenative N-N homocoupling reaction. Chem Commun 56:4348–4351. https://doi.org/10.1039/D0CC01052A
112. Kehl A, Gieshoff T, Schollmeyer D et al (2018) Electrochemical conversion of phthaldianilides to phthalazin-1,4-diones by dehydrogenative N-N bond formation. Chem A Eur J 24:590–593. https://doi.org/10.1002/chem.201705578

113. Khalil AM, Berghot MA, Gouda MA et al (2010) Synthesis and antibacterial studies of azodispersed dyes derived from 2-(thiazol-2-yl)phthalazine-1,4-diones. Monatsh Chem 141: 1353–1360. https://doi.org/10.1007/s00706-010-0392-3
114. Nematollahi J (1972) Reaction of alkylhydrazines. I. Reaction of 1,1-dimethylhydrazine with alkyl phthalates. J Heterocyclic Chem 9:963–965. https://doi.org/10.1002/jhet.5570090442
115. Correa A, Tellitu I, Domínguez E et al (2006) Novel alternative for the N-N bond formation through a PIFA-mediated oxidative cyclization and its application to the synthesis of indazol-3-ones. J Org Chem 71:3501–3505. https://doi.org/10.1021/jo060070
116. Lund H (1970) Electrolysis of N-heterocyclic compounds. In: Katritzky AR (ed) Advances in heterocyclic chemistry, vol 12. Elsevier, pp 213–316
117. Moinet C (2010) Electrochemistry of heterocycles. In: Bard AJ, Stratmann M (eds) Encyclopedia of electrochemistry. Wiley, pp 339–398
118. Liu Y, Li F, Li D et al (2023) Controllable synthesis of diazocine – investigation on electroreduction mechanism for intramolecular cyclization of 2,2′-dinitrodibenzyl in the presence of CO2. J Electroanal Chem 944:117644. https://doi.org/10.1016/j.jelechem.2023.117644
119. Huang B, Sun Z, Sun G (2022) Recent progress in cathodic reduction-enabled organic electrosynthesis: trends, challenges, and opportunities. eScience 2:243–277. https://doi.org/10.1016/j.esci.2022.04.006
120. Corsello MA, Kim J, Garg NK (2017) Indole diterpenoid natural products as the inspiration for new synthetic methods and strategies. Chem Sci 8:5836–5844. https://doi.org/10.1039/C7SC01248A
121. Umer SM, Solangi M, Khan KM et al (2022) Indole-containing natural products 2019-2022: isolations, reappraisals, syntheses, and biological activities. Molecules 27:7586. https://doi.org/10.3390/molecules27217586
122. Du P, Peters DG (2010) Reduction of 1-(2-Chloroethyl)-2-nitrobenzene and 1-(2-Bromoethyl)-2-nitrobenzene at carbon cathodes: Electrosynthetic routes to 1-Nitro-2-vinylbenzene and 1H-indole. J Electrochem Soc 157:F167–F172. https://doi.org/10.1149/1.3482019
123. Du P, Brosmer JL, Peters DG (2011) Electrosynthesis of substituted 1H-indoles from o-nitrostyrenes. Org Lett 13:4072–4075. https://doi.org/10.1021/ol2006049
124. Frontana-Uribe BA, Moinet C, Toupet L (1999) N-substituted 1-aminoindoles from electrogenerated N-substituted 2-(ortho-nitrosophenyl)ethylamines. Eur J Org Chem 1999: 419–430. https://doi.org/10.1002/(SICI)1099-0690(199902)1999:2<419:AID-EJOC419>3.0.CO;2-V
125. Lamoureux C, Moinet C, Tallec A (1986) Cellules d'electrolyse a circulation. montages particuliers et applications a des reactions electrochimiques consecutives et opposees. Electrochim Acta 31:1–12. https://doi.org/10.1016/0013-4686(86)80053-6
126. Frontana-Uribe BA, Moinet C (1998) 2-substituted indazoles from electrogenerated ortho-nitrosobenzylamines. Tetrahedron 54:3197–3206. https://doi.org/10.1016/S0040-4020(98)00058-1
127. Frontana-Uribe BA, Moinet C, Dory Y et al (1999) Synthesis and electrochemical behaviour of 2-N-substituted indazoles. Acta Chem Scand 53:814–823. https://doi.org/10.3891/acta.chem.scand.53-0814
128. Alberti A, Carloni P, Greci L et al (1991) Chemical and electrochemical reduction of ortho-nitroanilides. A combined chemical, polarographic and EPR study. J Chem Soc Perkin Trans 2:1019–1023. https://doi.org/10.1039/P29910001019
129. Sokolov AA, Syroeshkin MA, Solkan VN et al (2014) Efficient electrochemical synthesis of pyrido[1,2-a]benzimidazoles. Russ Chem Bull 63:372–380. https://doi.org/10.1007/s11172-014-0440-y
130. Sokolov AA, Begunov RS, Syroeshkin MA et al (2016) Electrochemical reduction of N-(2-nitro-4-R-phenyl)pyridinium salts using redox-mediators. Russ Chem Bull 65:209–214. https://doi.org/10.1007/s11172-016-1286-2

131. Herold S, Möhle S, Zirbes M et al (2016) Electrochemical amination of less-activated alkylated arenes using boron-doped diamond anodes. Eur J Org Chem 2016:1274–1278. https://doi.org/10.1002/ejoc.201600048
132. Möhle S, Herold S, Richter F et al (2017) Twofold electrochemical amination of naphthalene and related arenes. ChemElectroChem 4:2196–2210. https://doi.org/10.1002/celc.201700476
133. Elbs K, Keiper W (1903) Ein Verfahren Zur Darstellung von Phentriazolen. J Prakt Chem 67: 580–584. https://doi.org/10.1002/prac.19030670152
134. Hyo Kim B, Byung Lee D, Ho Kim D et al (2000) Electrochemical synthesis of 2-aryl-2H-benzotriazoles and their N-oxides by controlled potential cathodic electrolysis. Heterocycles 53:841–850. https://doi.org/10.3987/COM-99-8824
135. Wirtanen T, Rodrigo E, Waldvogel SR (2020) Selective and scalable electrosynthesis of 2H-2-(aryl)-benzod-1,2,3-triazoles and their N-oxides by using leaded bronze cathodes. Chem A Eur J 26:5592–5597. https://doi.org/10.1002/chem.201905874
136. Gault C, Moinet C (1989) Reaction of electrogenerated 2-nitrosobenzic acids with sulphinic acids. A convenient route to N-sulfonylbenzisoxazolones. Tetrahedron 45:3429–3436. https://doi.org/10.1016/S0040-4020(01)81021-8
137. Hyo Kim B, Moo Jun Y, Rack Choi Y et al (1998) Electrochemical synthesis of 2,1-benzisoxazoles by controlled potential cathodic electrolysis. Heterocycles 48:749–754. https://doi.org/10.3987/COM-97-8077
138. Hosseini S, Bawel SA, Mubarak MS et al (2019) Rapid and high-yield electrosynthesis of benzisoxazole and some derivatives. ChemElectroChem 6:4318–4324. https://doi.org/10.1002/celc.201801321
139. Wang L-W, Feng Y-F, Lin H-M et al (2021) Electrochemically enabled selenium catalytic synthesis of 2,1-benzoxazoles from o-nitrophenylacetylenes. J Org Chem 86:16121–16127. https://doi.org/10.1021/acs.joc.1c00012
140. Rodrigo E, Baunis H, Suna E et al (2019) Simple and scalable electrochemical synthesis of 2,1-benzisoxazoles and quinoline N-oxides. Chem Commun 55:12255–12258. https://doi.org/10.1039/c9cc06054e
141. Baumann M, Baxendale IR (2013) An overview of the synthetic routes to the best selling drugs containing 6-membered heterocycles. Beilstein J Org Chem 9:2265–2319. https://doi.org/10.3762/bjoc.9.265
142. Lund H, Feoktistov LG, Nielsen PH (1969) Electroorganic preparations. XXVIII. Preparation of some heteroaromatic N-oxides by ring closure. Acta Chem Scand 23:3482–3492. https://doi.org/10.3891/acta.chem.scand.23-3482
143. Winter J, Prenzel T, Wirtanen T et al (2023) Direct electrochemical synthesis of 2,3-disubstituted quinoline N-oxides by cathodic reduction of nitro arenes. Chem A Eur J 29:e202203319. https://doi.org/10.1002/chem.202203319
144. Zhao H-B, Xu P, Song J et al (2018) Cathode material determines product selectivity for electrochemical C-H functionalization of Biaryl Ketoximes. Angew Chem Int Ed 57:15153–15156. https://doi.org/10.1002/anie.201809679
145. Chibani A, Hazard R, Tallec A (1991) Electrochemical reduction of N-(o-nitrobenzoyl) and N-(o-nitrobenzyl)-amides or imides. Electrosynthesis of quinazoline derivatives. Bull Soc Chim Fr:814–822
146. Estrada E, Rieker A (1994) Electrosynthesis of 2-Alkyl-4(3H)-quinazolinones. Z Naturforsch B 49:1566–1568. https://doi.org/10.1515/znb-1994-1119
147. Koleda O, Prenzel T, Winter J et al (2023) Simple and scalable electrosynthesis of 1H-1-hydroxy-quinazolin-4-ones. Chem Sci 14:2669–2675. https://doi.org/10.1039/D3SC00266G
148. Sicker D, Hartenstein H, Hazard R et al (1994) Syntheses for 2-hydroxy-2 H-1,4-benzothiazin-3(4H)-one derivatives as thio analogues of natural hemiacetals. J Heterocyclic Chem 31:809–812. https://doi.org/10.1016/0013-4686(95)00083-Q
149. Sicker D, Hartenstein H, Mouats C et al (1995) Electrochemical reduction of o-nitrophenylthioacetic derivatives. Production of 2H-1,4-benzothiazines. Electrochim Acta 40:1669–1674. https://doi.org/10.1016/0013-4686(95)00083-Q

150. Matschiner H, Tanneberg H, Maschmeier C-P (1981) Elektrosynthese cyclischer Hydroxamsäuren mit anschließender analoger Bamberger-Umlagerung. J Prakt Chem 323: 924–926. https://doi.org/10.1002/prac.19813230610
151. Mouats C, Hazard R, Raoult E et al (1994) Electrochemical reduction of o-nitrophenoxy compounds: an access to 2H-1, 4-benzoxazine derivatives. Bull Soc Chim Fr 131:71–77
152. Winter J, Prenzel T, Wirtanen T et al (2024) Highly selective scalable electrosynthesis of 4-hydroxybenzo[e]-1,2,4-thiadiazine-1,1-dioxides. Cell Rep Phys Sci:101927. https://doi.org/10.1016/j.xcrp.2024.101927
153. Cantillo D (2022) Synthesis of active pharmaceutical ingredients using electrochemical methods: keys to improve sustainability. Chem Commun 58:619–628. https://doi.org/10.1039/d1cc06296d
154. Hatch CE, Chain WJ (2023) Electrochemically enabled total syntheses of natural products. ChemElectroChem 10:e202300140. https://doi.org/10.1002/celc.202300140
155. Geske L, Sato E, Opatz T (2020) Anodic oxidation as an enabling tool for the synthesis of natural products. Synthesis 52:2781–2794. https://doi.org/10.1055/s-0040-1707154
156. Kingston C, Palkowitz MD, Takahira Y et al (2020) A survival guide for the "Electro-curious". Acc Chem Res 53:72–83. https://doi.org/10.1021/acs.accounts.9b00539
157. Gütz C, Klöckner B, Waldvogel SR (2016) Electrochemical screening for electroorganic synthesis. Org Process Res Dev 20:26–32. https://doi.org/10.1021/acs.oprd.5b00377
158. Hielscher MM, Dörr M, Schneider J et al (2023) LABS: laboratory automation and batch scheduling – a modular open source python program for the control of automated electrochemical synthesis with a web Interface. Chem Asian J 18:e202300380. https://doi.org/10.5445/IR/1000168377
159. Rein J, Annand JR, Wismer MK et al (2021) Unlocking the potential of high-throughput experimentation for electrochemistry with a standardized microscale reactor. ACS Cent Sci 7: 1347–1355. https://doi.org/10.1021/acscentsci.1c00328
160. Rial-Rodríguez E, Williams JD, Cantillo D et al (2024) An automated electrochemical flow platform to accelerate library synthesis and reaction optimization. ChemRxiv. https://doi.org/10.26434/chemrxiv-2024-dm4zj
161. Seidler J, Strugatchi J, Gärtner T et al (2020) Does electrifying organic synthesis pay off? The energy efficiency of electro-organic conversions. MRS Energy Sustain 7:42. https://doi.org/10.1557/mre.2020.42
162. Lehnherr D, Chen L (2024) Overview of recent scale-ups in organic electrosynthesis (2000–2023). Org Process Res 28:338–366. https://doi.org/10.1021/acs.oprd.3c00340
163. Leech MC, Garcia AD, Petti A et al (2020) Organic electrosynthesis: from academia to industry. React Chem Eng 5:977–990. https://doi.org/10.1039/D0RE00064G
164. Cardoso DSP, Šljukić B, Santos DMF et al (2017) Organic Electrosynthesis: from laboratorial practice to industrial applications. Org Process Res Dev 21:1213–1226. https://doi.org/10.1021/acs.oprd.7b00004

Top Heterocycl Chem (2024) 61: 91–112
https://doi.org/10.1007/7081_2024_81

Published online: 20 September 2024

Electrochemical Dearomatization of Indoles: Access to Diversified Functionalized Spirocyclic Indolines

Jian Wang, Ranran Zhu, Yufen Zhao, and Ju Wu

Contents

Abstract Spirocyclic indolines are widely present in drug-like molecules and alkaloids, so the synthesis of functionalized spirocyclic indolines could provide a promising approach to enrich the drug candidate library, which is highly important in drug discovery. Radical-induced dearomatization of indole can offer a straightforward strategy to access the dearomatized skeleton, which has gained increased significance in the past decades. In contrast to conventional organic synthesis, electrosynthesis has become a green and sustainable strategy for the oxidation of indole nucleus and for the generation of radicals. This review will focus on the

J. Wang and J. Wu (✉)
Institute of Drug Discovery Technology, Ningbo University, Ningbo, Zhejiang, China
e-mail: wuju@nbu.edu.cn

R. Zhu
Institute of Drug Discovery Technology, Ningbo University, Ningbo, Zhejiang, China

Institute of Mass Spectrometry, School of Material Science and Chemical Engineering, Ningbo University, Ningbo, Zhejiang, China

Y. Zhao
Institute of Drug Discovery Technology, Ningbo University, Ningbo, Zhejiang, China

College of Chemistry and Chemical Engineering, Xiamen University, Xiamen, Fujian, China

electrochemical dearomatization of indoles for the synthesis of diversified, functionalized 3,3-spirocyclic indolines over the past decade.

Keywords Dearomatization · Electrosynthesis · Radical cyclization · Spirocyclic indoline

1 Introduction

Spiro heterocycles are privileged structural motifs, which are widely present in a series of bioactive compounds and natural products. The tetrahedral nature of the spirocyclic carbon results in conformational restriction and structural complexity, which confers on the motif more diverse biological activity and pharmacological properties [1, 2]. Nowadays, the incorporation of spiro building blocks into more molecules with pharmacological applications, such as antidiabetic, anticancer, anti-Alzheimer's disease, and so on, has been proven effective [3, 4]. In particular, the synthesis of spirocyclic indoline derivatives, which are an important class of spirocyclic compounds present in a wide range of pharmaceuticals and biologically important natural alkaloids, has attracted considerable attention [5–8]. In spite of some elegant and remarkable achievements made in the past several decades, general and efficient protocols to access a diverse array of functionalized indolines are still of great importance (Scheme 1).

Dearomatization of indoles can deliver the functionalized indolines with two contiguous stereogenic centers, which is of high interest for the synthesis of drug-like compounds and natural products. Since it can provide a straightforward approach to converting the planar indoles to three-dimensional skeletons, considerable efforts have been devoted to the dearomative difunctionalization of indoles. In this regard, radical-induced dearomatization of indoles has gained more and more attention from synthetic chemists. Given the usage of stoichiometric redox agents, which are traditionally used to generate the radicals from the corresponding precursors, the atom efficiency and sustainability of these reactions have significant opportunities for improvement.

The first organic electrosynthesis can be traced back to the first half of the nineteenth century, achieved by Faraday and Kolbe groups [9, 10]. It emerged somewhat later as an important synthetic method in the laboratory and has now occupied such a position for nearly a hundred years. Along with the development of

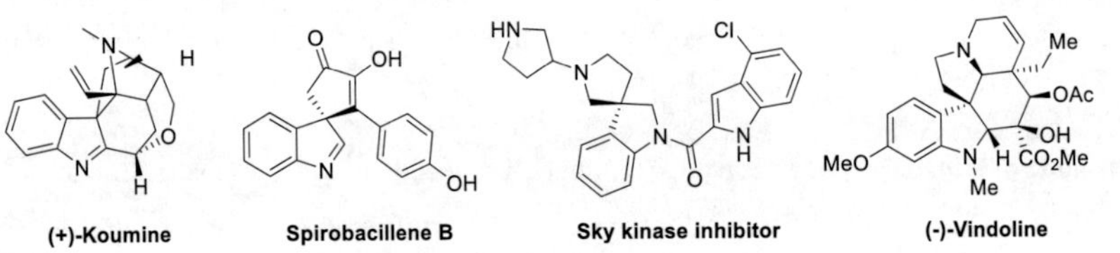

Scheme 1 Selected examples of spirocyclic indolines

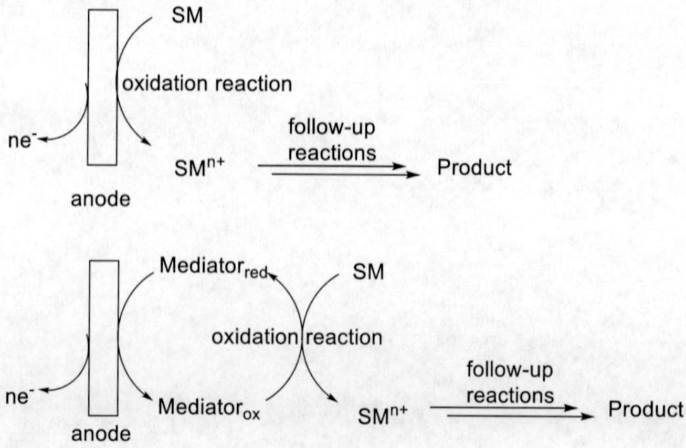

Scheme 2 Electrochemical direct oxidation and indirect oxidation

electrochemical theory and apparatuses, electrosynthesis has been applied widely, including by industry for a selection of bulk chemical productions over the past several decades [11]. For example, paired electrosynthesis was adopted by BASF for the simultaneous production of 4-*tert*-butylbenzaldehyde dimethylacetal and phthalide in the undivided cell. Pairing an oxidative synthesis with a complementary reductive one was beneficial for preserving resources and minimizing waste [12]. Electrosynthesis can effectively avoid the usage of redox agents, making the reaction safer and more efficient. Also, the redox agent is replaced by traceless electrons in electrochemical conditions, which can effectively reduce side reactions and separation difficulties. Given its innate sustainability, scalability, and versatility, electrochemistry is a promising technology for green organic synthesis.

Mechanistically, there are two types of electrochemical oxidation: direct electrolytic oxidation and indirect oxidation. In the first case, the reactants lose electrons directly on the surface of the anode, leading to the formation of the activated species in situ, which can be further converted to the final product. In contrast, indirect oxidation is carried out in the presence of a redox mediator. The electrochemically activated mediator formed on the surface of the anode can oxidize the reactants in the solution by a homogeneous redox reaction [13] (Scheme 2). In this review, we will focus on electrochemical synthesis of diversified functionalized spirocyclic indolines from indole starting materials.

2 Electrochemical Synthesis of Functionalized Spirocyclic Indolines

Traditionally, the innately nucleophilic indole derivatives have been dearomatized to indolines through reactions with a series of electrophiles. On the other hand, radical-mediated electrochemical dearomatization of indoles has attracted increasing

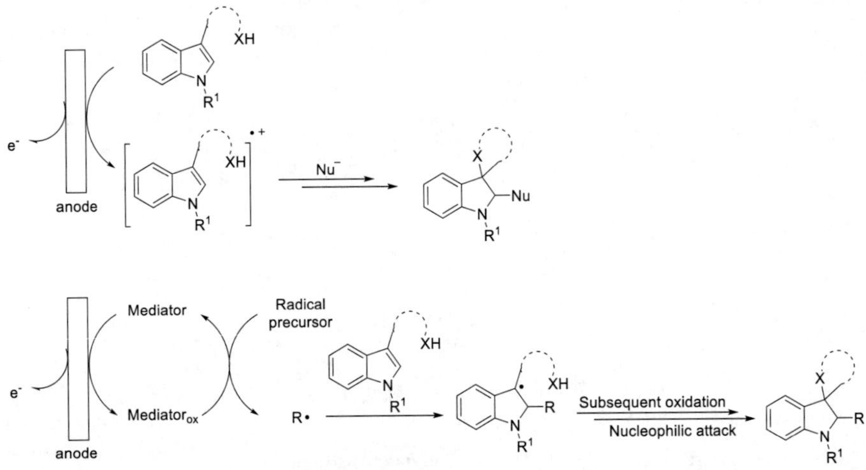

Scheme 3 Electrochemical dearomatization of indole derivatives

attention from synthetic community over the past decade. In general, two main modes of electrochemical dearomatization of indoles can be distinguished. The first involves cross-coupling reactions between indolyl radical cations formed by anodic oxidation with various nucleophiles. The other involves introducing a variety of electrochemically generated radicals to the double bond of indole substrates (Scheme 3).

2.1 Synthesis of Methoxylated Spirocyclic Indolines

In 2019, the Vincent group developed an electrochemical dearomative 2,3-difunctionalization of indoles, leading to 2,3-dialkoxy or 2,3-diazido indolines **2** in the undivided cell [14]. This operationally simple method can install two C-O or C-N bonds simultaneously into the indole nucleus under external-oxidant-free conditions. A wide range of indole substrates have been proven suitable to generate the *trans*-products as the major diastereoisomers with good to excellent yields. Noteworthily, the methoxylated spirocyclic indolines can be furnished in moderate yields and excellent diastereoselectivity when the indole substrates **1** with a tosylamide-containing C3-side chain were subject to the same conditions. Interestingly, the major products **2a-e** were methoxylated tetrahydropyranoindolines, instead of the spirocyclic products **2f-j**, when the alcohol served to attack the carbocation intermediate. On the contrary, the dimethoxylated indolines were obtained in acceptable yields from tryptophan or tryptophol derivatives, leaving the nucleophilic motif on the side chain intact (Scheme 4).

Moreover, a radical clock experiment was performed under the optimal conditions, and the desired dearomatized product **2k** was afforded in 83% yield without

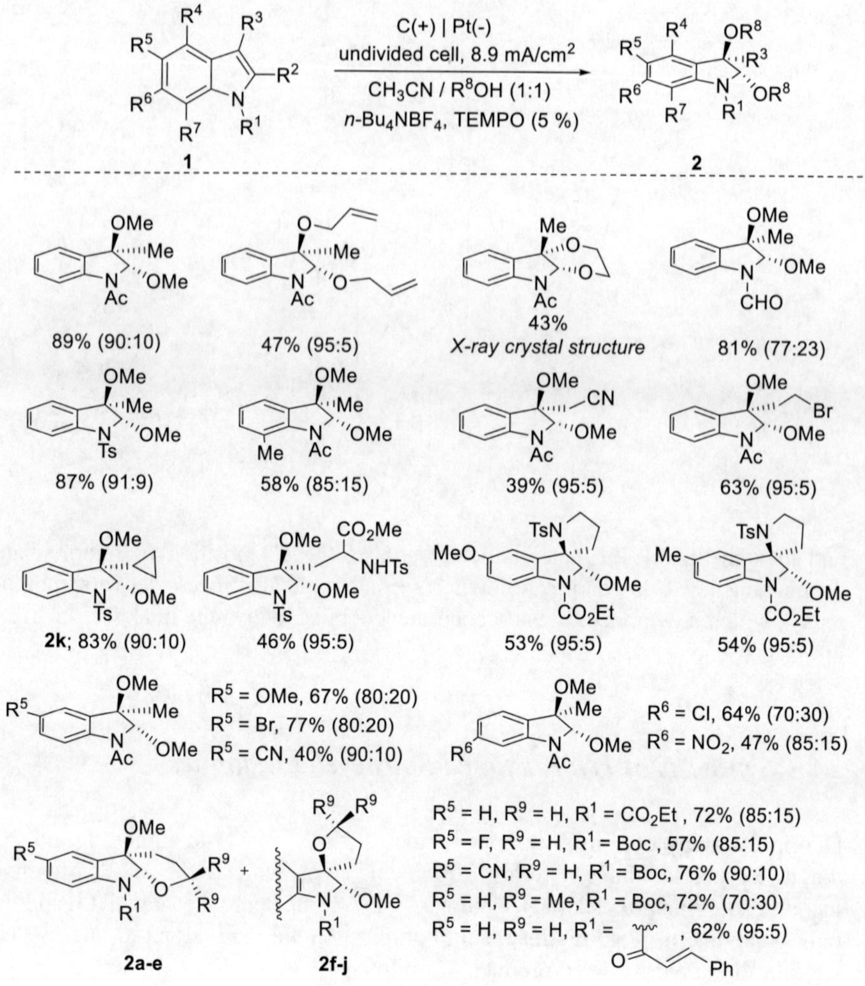

Scheme 4 Electrochemical 2,3-difunctionalization of indoles

opening of the cyclopropyl group. This experiment demonstrated that a radical pathway is less likely to be involved in this case. Since there is no catalytic current detected when TEMPO was added to a solution of *N*-Ac 3-methylindole, TEMPO is unlikely to act as a mediator in this electrochemical process. Furthermore, the oxidation potential of *N*-Ac 3-methylindole is 1.6 V (vs Ag/AgCl), which is higher than that of TEMPO (0.7 V vs Ag/AgCl). The reaction was also performed at a constant anode potential in the presence of reference electrode (1.3 V vs Ag/AgCl), which failed to generate the methoxylated product. This outcome indicated that the reaction mainly results from oxidation of the indole nucleus, instead of TEMPO. Overall, the proposed mechanism involves indolyl radical cations **3**. First, indole substrates **1** are converted to indolyl radical cations **3** in a direct anodic oxidation,

Scheme 5 Mechanistic proposal for electrochemical 2,3-difunctionalization of indoles

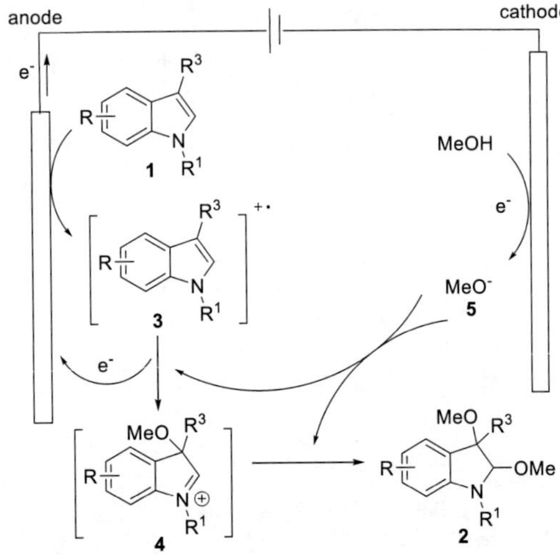

and then methoxide **5** nucleophilically attacks the C3 position of intermediate, forming the new C-O bond. A following, second oxidation process delivers iminium ions **4**, which are trapped by the second nucleophiles to give the final dearomatized products (Scheme 5).

2.2 *Synthesis of Hydroxylated Spirocyclic Indolines*

During the study of substrate scope in the previous work, the authors found that dihydroxylated indolines were formed in some cases, which indicated that hydroxide, instead of methoxide or azide anions, is also competent to react with the indolyl radical cations. Inspired by this result, optimization was carried out to improve the yield of the dihydroxylated product. Unfortunately, the yields were not acceptable even though the reaction was performed in a mixture of H_2O and CH_3CN. This result demonstrated that the direct electrochemical oxidation of the indole ring might not provide an efficient path to the dihydroxylated product in this case, indicating that a new type of electrochemical indole activation is highly desired.

In 2020, a simple and sustainable electrochemical dearomative dihydroxylation and hydroxycyclization of indoles was reported by the Vincent group [15]. In this work, $MgBr_2$ was used as a redox mediator to oxidize the indole nucleus into a bromonium ion, which can be converted to the dihydroxylated indolines in the presence of H_2O. Remarkably, the indole substrates **6** with different nucleophilic entities on the C3-side chain have a significant effect on the products. For example, the alcohol on the side chain could attack the C2 position of the indole ring, leading to the hydroxycyclization product **7**, without the formation of the hydroxylated spirocyclic indolines. On the contrary, the tosylamide derivatives **6** can yield the

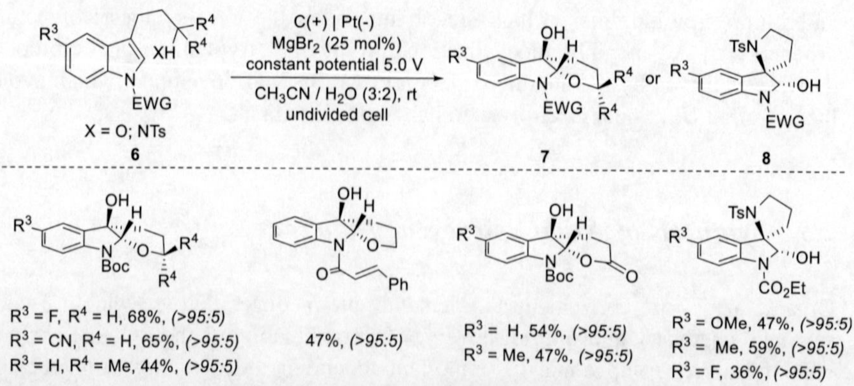

Scheme 6 Electrochemical dearomative hydroxycyclization of indoles

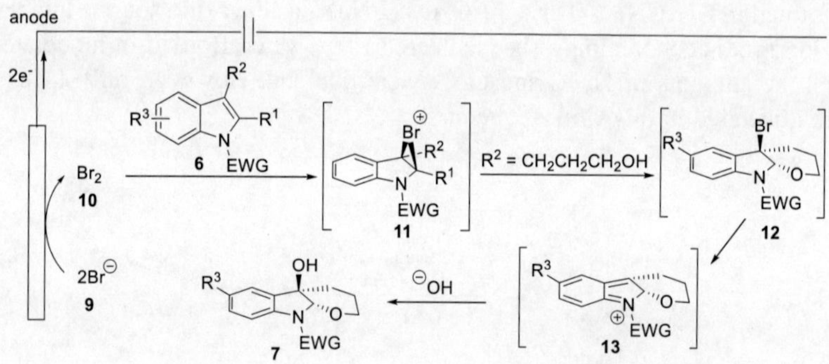

Scheme 7 Mechanistic proposal for electrochemical hydroxycyclization of indoles

corresponding spirocyclic products **8** with an excellent diastereoselectivity (Scheme 6).

To gain more insight into the mechanism, control experiments were carried out by the authors. The reaction proceeded smoothly to generate the dihydroxylated indolines in the presence of an electrophilic-bromine reagent in nonelectrochemical conditions, implying that a bromonium ion intermediate could be involved in this process. Moreover, a sharp decrease in the yield was found when the reaction was performed in the absence of $MgBr_2$, even though a higher voltage (5 V cell potential) was applied. Therefore, direct oxidation of the indole nucleus is likely not involved in the dominant pathway. Given the results of cyclic voltammetry experiments, the authors proposed that the major pathway involves a two-electron anodic oxidation of two bromides **9** into molecular bromine (**10**), which could react with the double bond of the indole ring to furnish the key bromonium ion intermediate **11** (Scheme 7).

Notably, the same mechanism has been adopted by the Vincent group in the electrochemical synthesis of 3a-bromofuranoindolines and

3a-bromopyrroloindolines, which are of interest in the total synthesis of natural products [16]. A general bromocyclization of tryptophol, tryptamine, and tryptophan derivatives driven by the anodic oxidation of $MgBr_2$ was developed, which avoids the use of an additional electrolyte and an external oxidant.

2.3 Synthesis of Azido Spirocyclic Indolines

Organic azides are convenient and versatile intermediates that are widely used in synthetic chemistry and drug discovery. The combination of the indoline skeleton with the azide group could be important to enrich the diversity of spirocyclic indoline derivatives. In 2014, the Shi group reported the oxidative azidation of the indole nucleus followed by a ring closure in the presence of ceric ammonium nitrate (CAN), delivering the spirocyclic 2-azido indolines **15** [17]. The reaction was performed in CH_3CN, at 0 or −40°C, using NaN_3 as the azido source. Interestingly, diazido products containing six-membered rings were afforded in moderate yields when the amounts of NaN_3 and CAN were doubled. However, only 11 examples were illustrated in this work (Scheme 8).

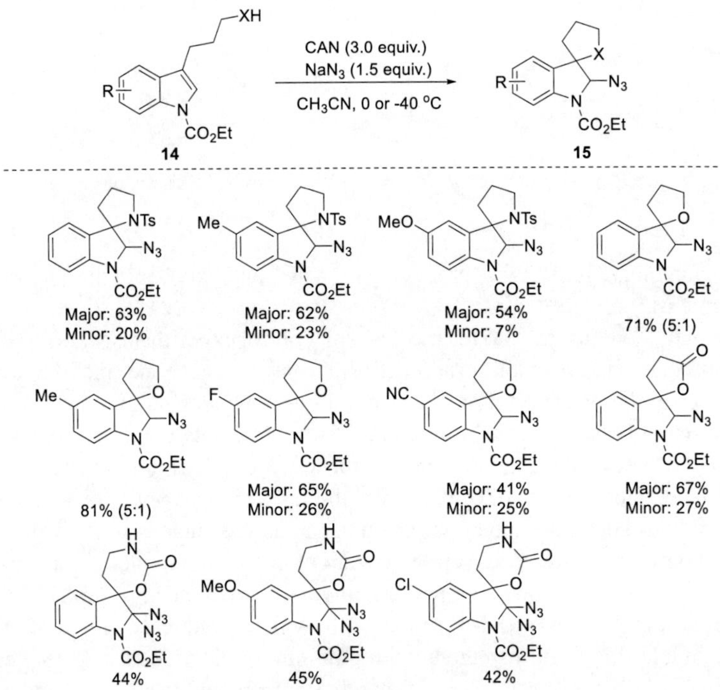

Scheme 8 CAN-promoted synthesis of azido spirocyclic indolines

To address the stoichiometric CAN used in the previous work, which poses a dramatic challenge to green chemistry, an efficient and environmentally friendly electrochemical protocol is still of great importance. In 2022, our group developed an electrochemical synthesis of azido-containing spirocyclic indolines mediated by $MnBr_2$ using a constant cell potential [18]. The indole substrates **16** bearing either electron-donating groups or electron-withdrawing groups proved capable of generating the corresponding products **17** in acceptable to good yields with moderate diastereoselectivity. Moreover, pendant nucleophiles including alcohol, carboxylic acid, or tosylamide motifs were well tolerated under the electrochemical conditions (Scheme 9).

On the basis of previous work and their mechanistic study, the authors proposed that the reaction was driven by the electrochemical oxidation of Mn^{II}-N_3 species **18**, formed in a ligand exchange between $MnBr_2$ and azide anion, which was supported

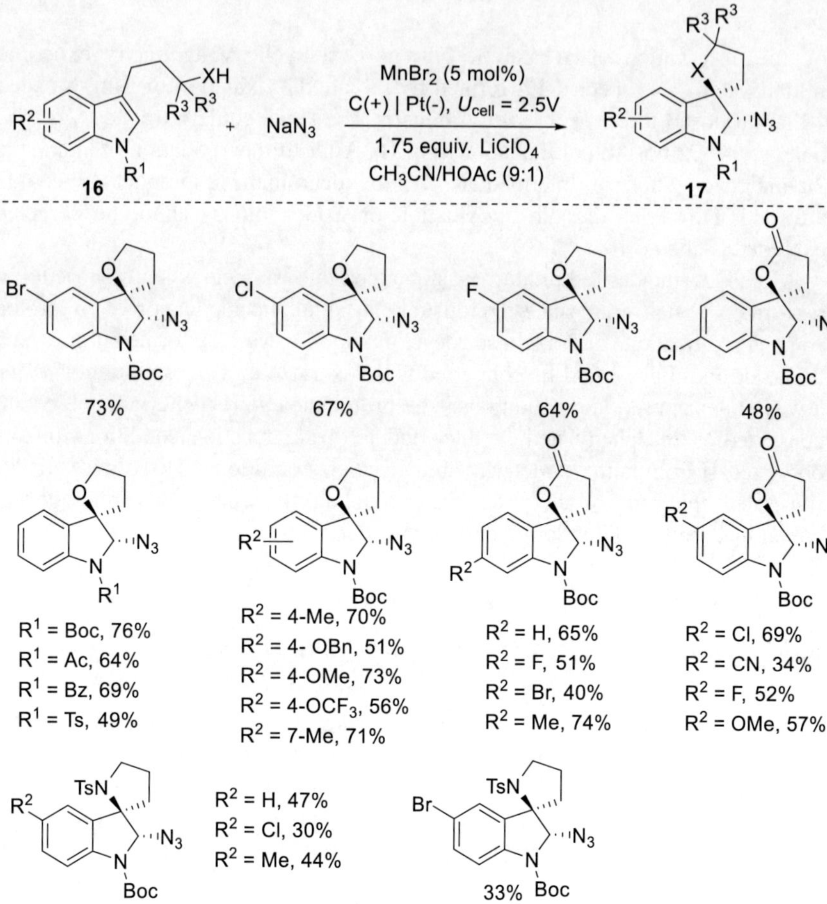

Scheme 9 Electrochemical synthesis of azido spirocyclic indolines

Scheme 10 Mechanistic proposal for electrochemical synthesis of azido spirocyclic indolines

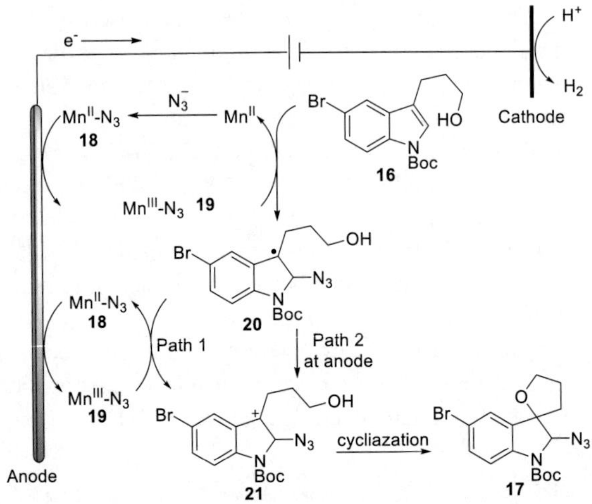

by the observation of a catalytic current in a cyclic voltammetry experiment. Initially, Mn^{III}-N_3 species **19**, formed by the anodic oxidation of Mn^{II}-N_3 species **18**, is proposed to act as an azodyl transfer agent that could introduce a new C-N bond at the C2 position of the indole ring **16**. After further oxidation of intermediate **20** and cyclization of intermediate **21**, the dearomatized product **17** could be afforded. However, the direct oxidation of azide anions cannot be completely excluded (Scheme 10).

Moreover, the desired product was obtained in 67% yield when the reaction was performed on half-gram scale. To illustrate the synthetic application of our protocol, some derivatizations of 2-azido spirocyclic indolines were performed. For example, the azido indolines could be converted to 1,2,3-triazole or phosphoramidate derivatives under appropriate conditions. The brominated spirocyclic indoline could be converted to the phenylated product under Suzuki reaction conditions or to the borylated spiroindoline under Miyaura reaction conditions. These derivatizations illustrated that our protocol can be employed to access diversified nitrogen-containing spirocyclic indoline derivatives (Scheme 11).

2.4 *Synthesis of Fluorinated Spirocyclic Indolines*

Since the incorporation of fluorine in drug-like molecules can increase their bioavailability, chemical and metabolic stability, and protein-binding affinity, fluorine-containing groups are widely present in pharmaceuticals and agrochemicals. In 2017, the Vincent group developed a radical-mediated dearomatization of indoles for the synthesis of fluoroalkylated spirocyclic indolines **23**. A broad substrate scope and excellent diastereoselectivity were illustrated [19]. The indole double bond is attacked by a trifluoromethyl radical, formed from the Langlois' reagent by 3.0

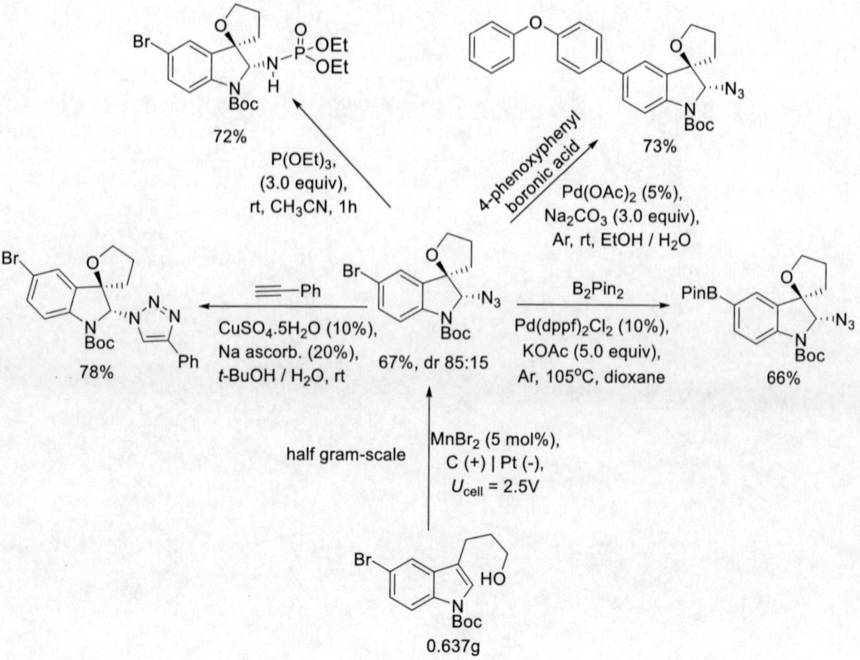

Scheme 11 Synthetic applications of spirocyclic indoline

equiv. of CAN, delivering the intermediate with a radical on the carbon adjacent to the newly formed bond. Then, a carbocation intermediate could be formed by single-electron transfer (SET) in the presence of CAN, and the carbocation can be trapped by a nucleophilic entity on the C3 position, leading to the final products. Interestingly, this approach was also able to introduce a 1,1-difluoroethyl group in lieu of the trifluoromethyl group when $NaSO_2CF_2Me$ was used as the radical source (Scheme 12).

In the same year, the Yu group reported a practical strategy for catalytic trifluoromethylative dearomatization of indoles **24** and furans **26** under a CO_2 atmosphere via copper catalysis, leading to a variety of CF_3-containing spirocyclic indolines **25** and spiroacetals **27** in moderate to good yields [20]. The reaction was performed in 1 atm CO_2 in the presence of 2.0 equiv. of Togni's reagent and catalyzed by $[Cu(MeCN)_4]PF_4$. Notably, the azaindole derivatives, which rarely have been applied in dearomatizations, proved capable of generating the corresponding products. To shed light on the catalytic process, density functional theory (DFT) calculations were applied, which revealed that this transformation includes deprotonation of the amine, insertion of CO_2 into the Cu–N bond, activation of the Togni's reagent, addition of trifluoromethyl radical to indole, and intramolecular C–O coupling (Scheme 13).

Since nontoxic visible light can reduce the need for harsh redox reagents, photocatalysis has become increasingly popular in recent years. In 2019, the Gianetti

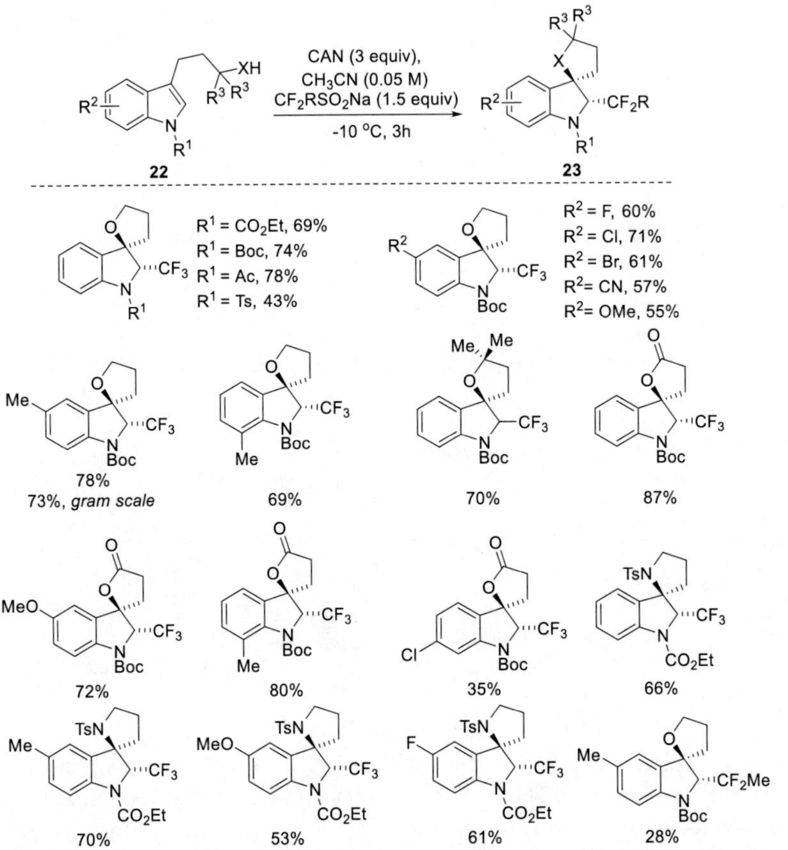

Scheme 12 CAN promoted synthesis of fluorinated spirocyclic indolines

group reported a red light-mediated cascade intramolecular trifluoromethylative dearomatization of indoles **28** with Umemoto's reagent, demonstrating a broad substrate scope and excellent diastereoselectivity [21]. Moreover, the six-membered heterocyclic indoline **29** could be afforded in a good yield when a longer carbon tether on the C3 position of indole was used to trap the carbocation intermediate. Remarkably, the helical carbenium ion (^{n}Pr-DMQA^{+}) was used as the organic photoredox catalyst, which could reduce Umemoto's reagent to generate CF_3 radical through oxidative quenching. Additionally, red light is low-energy, safer, and more abundant from sunlight, which could render the reaction conditions greener and milder (Scheme 14).

Based on experimental results and previous studies, a red light-induced dearomative process mediated by ^{n}Pr-DMQA^{+} **30** was proposed by the authors. First, under red light irradiation, the organic catalyst could be exited to ^{n}Pr-DMQA^{+*} **31**, which was able to generate the CF_3 radical from Umemoto's reagent. The

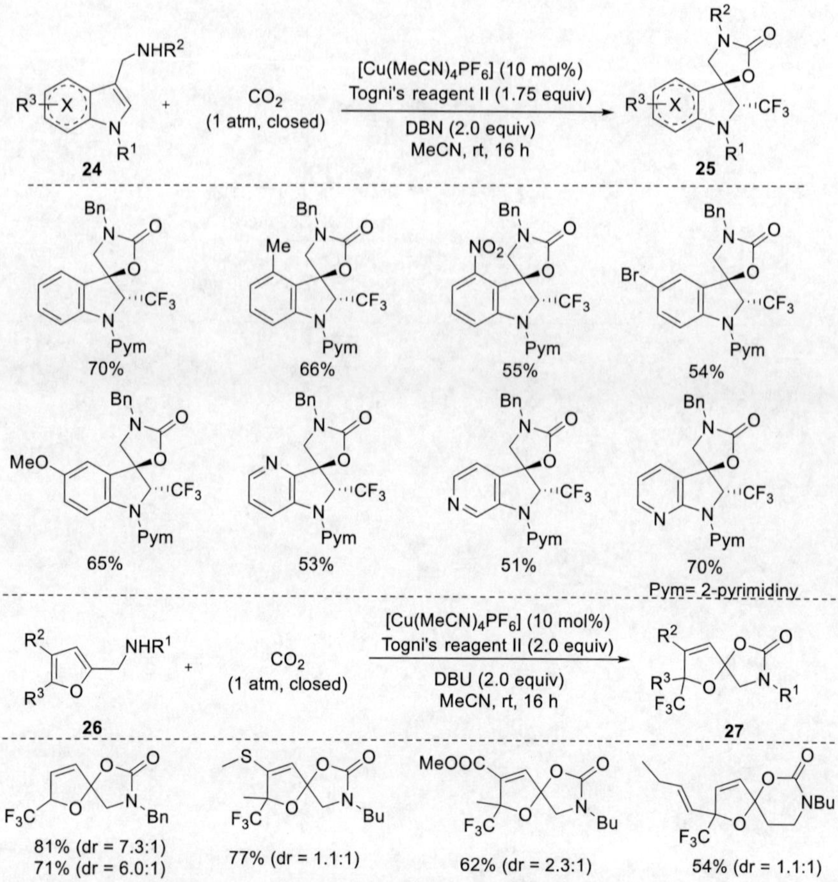

Scheme 13 Trifluoromethylative dearomatization of indoles and furans

resulting intermediate **33**, formed by radical addition to indole nucleus, could be further oxidized in the presence of ^{n}Pr-DMQA^{++} **32**, leading to the key carbocation species **34** that could form a new ring on the C3 position. Ground-state ^{n}Pr-DMQA^{+} could be regenerated during the oxidation of **33** to **34** and involved in the next catalytic cycle (Scheme 15). Furthermore, the steric hinderance between the CF_3 and the pendant carboxylic acid groups of two possible conformations of the carbocation intermediate could account for the excellent diastereoselectivities.

Given the price of the Umemoto's reagent, a general and sustainable protocol from cost-effective reagents is still attractive. In 2022, our group developed an electrochemical version that employs an affordable, stable, and efficient Langlois' reagent as the fluoroalkyl source and without the usage of external oxidants [22]. In this case, the reaction was carried out in an undivided cell and mediated by catalytic amount of $MnBr_2$ at a constant cell potential. The compatibility of pendant

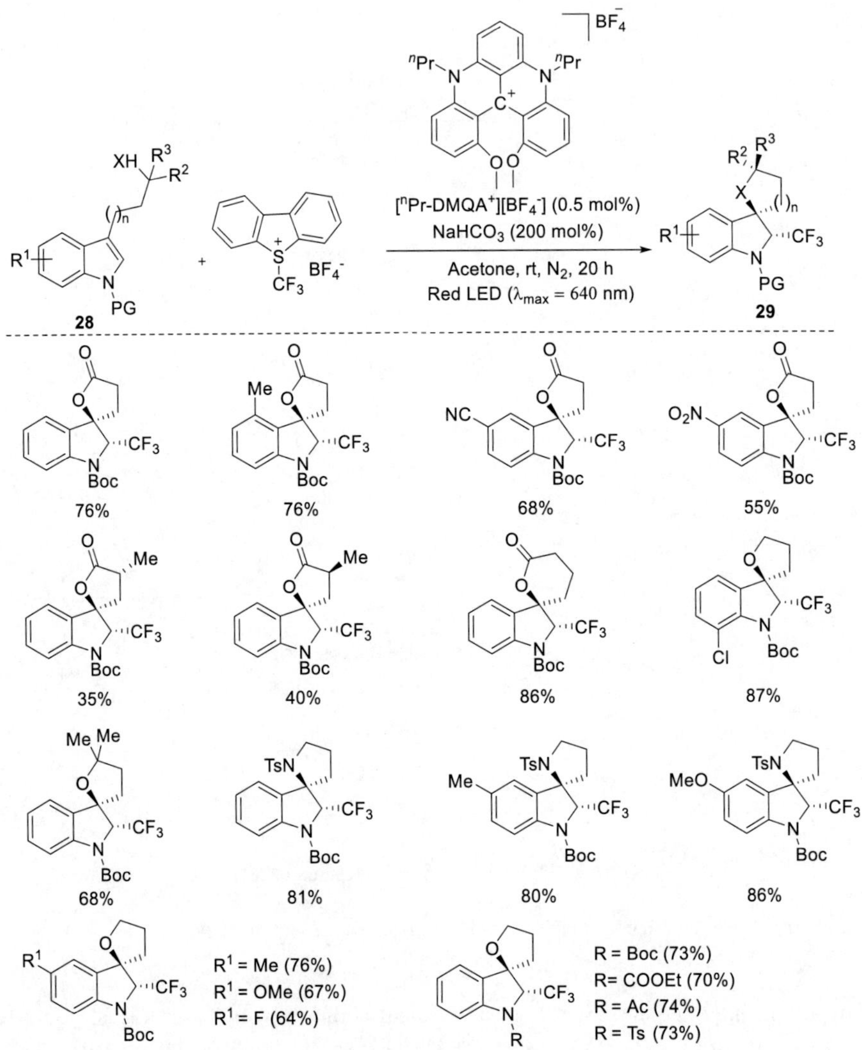

Scheme 14 Red light-mediated trifluoromethylation of indole derivatives

nucleophiles on the indole ring was evaluated; the corresponding products **36** were obtained in moderate yields. Notably, this approach was also capable of generating difluoromethylated 3,3-spiroindolines when $NaSO_2CF_2H$ was used under the same conditions. Interestingly, the desired product could be obtained in a 62% yield when 1.2 g of the corresponding indole substrate was subjected to the electrochemical conditions (Scheme 16).

Based on the results of cyclic voltammetry experiments, $NaSO_2CF_3$ (0.75 V vs SCE), bromide ions (0.85 V vs SCE), and the Mn^{II} species (1.10 V vs SCE) are all oxidized at more modest potentials than indole nucleus **35**, which avoids direct

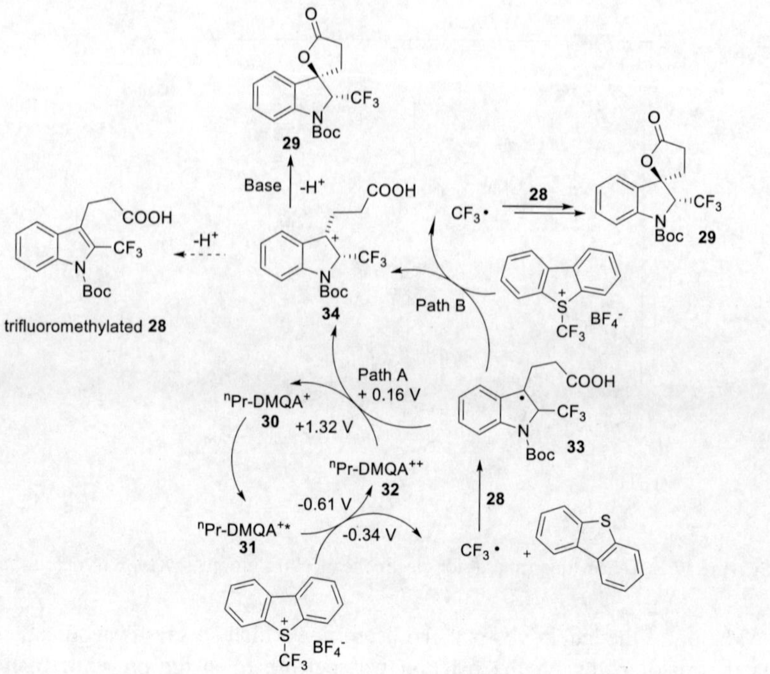

Scheme 15 Mechanistic proposal for red light-mediated trifluoromethylation of indoles

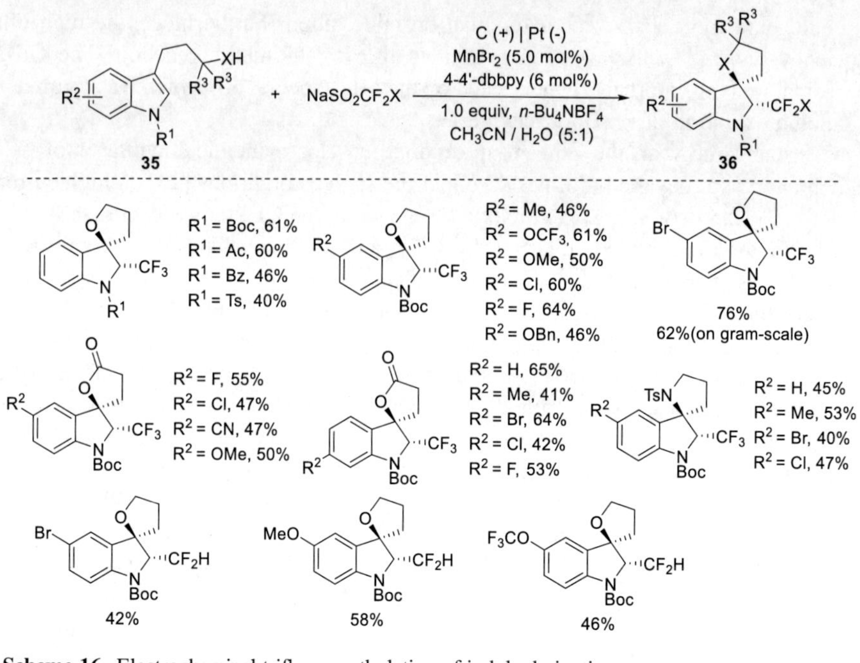

Scheme 16 Electrochemical trifluoromethylation of indole derivatives

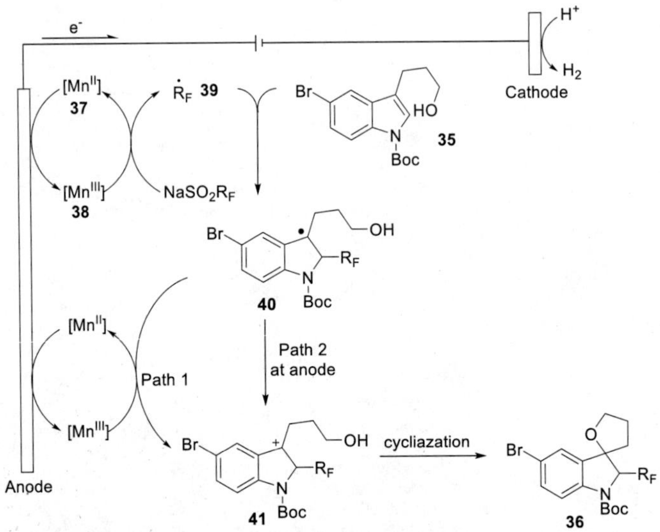

Scheme 17 Mechanistic proposal for electrochemical trifluoromethylation of indoles

oxidation of the latter substrate. To determine which species is anodically oxidized in the major pathway, the reaction was performed in the presence of n-Bu_4NBr instead of $MnBr_2$, leading to the desired product in a lower yield. In contrast, a similar yield was obtained when the reaction was performed under the same electrochemical conditions but mediated by $MnCl_2$, demonstrating that Mn^{II} species **37** could be a more active mediator than bromide anion. Furthermore, the high efficiency of Mn^{III} salts has been proven in the stoichiometric reaction. The CF_2X radical was thus more likely generated by the Mn^{III} species **38**, which was formed by anodic oxidation of $MnBr_2$ (Scheme 17).

In the same year, the Zhu group reported an electrochemical synthesis of CF_3-containing spirocyclic indolines **43** under metal-free conditions [23]. A high current density (12.5 mA/cm^2) was employed to generate the CF_3 radical from $NaSO_2CF_3$ directly in an undivided cell. Noteworthily, an indole substrate bearing an estrone motif was well tolerated, leading to the corresponding product in a 51% yield. Interestingly, the benzofuran substrate was converted to the desired dioxaspirononane in a moderate yield under the similar conditions. Other fluorine-containing reagents like HCF_2SO_2Na or $CF_3CF_2SO_2Na$ were also evaluated in the optimal conditions, but unfortunately, no desired spirocyclic products were afforded. In contrast to our work, indole substrates with an N-Boc protecting group proved incompatible in this direct anodic-oxidation protocol (Scheme 18).

To evaluate the anticancer bioactivity of the CF_3-containing spirocyclic indolines and indole derivatives, some products were tested in vitro against a panel of cancerous and normal cells, including liver cancer cell line Huh-7, lung cancer cell line A549, cisplatin (DDP)-resistant cell line A549/DDP, and human embryonic kidney epithelial cell line HEK293 T (normal cells) based on antiproliferation

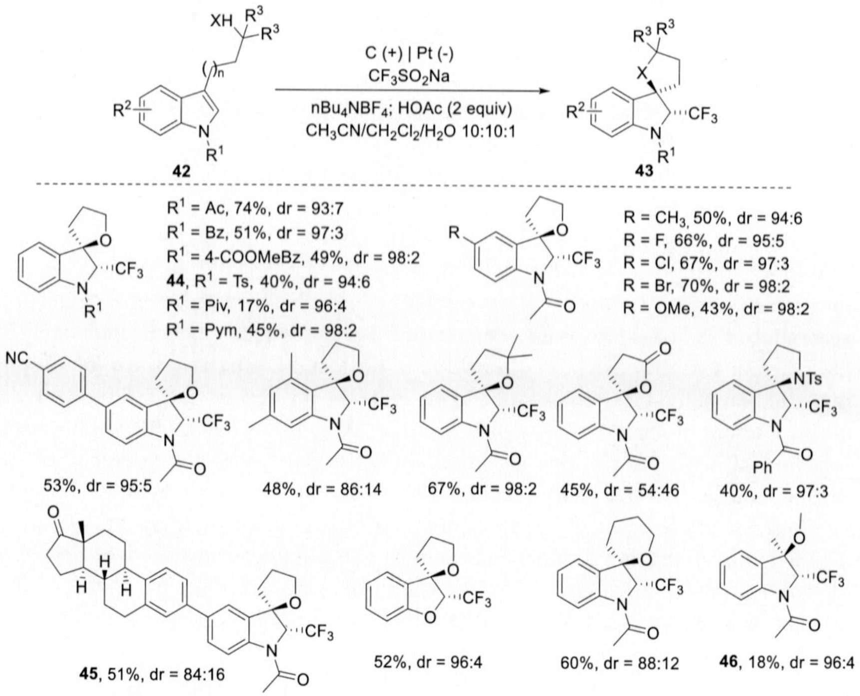

Scheme 18 Electrochemical trifluoromethylated indole derivatives

Table 1 IC50 (μM) values of compounds toward cancer cells

IC_{50}(μM)				
compound ID	A549-DDP	A549	Huh-7	HEK293T
44	13.7	>200	>200	>200
45	17.4	>200	1.1	>200
46	14.0	–	–	–

assays. The author has found that spirocyclic indolines, **44**, **45**, and **46** have a stronger anticancer activity than indole derivatives. Furthermore, the favorable anticancer activity in liver cancer cells of the Huh-7 cell line with IC_{50} values of 1.1 μM was observed in the case of compound **46**, which could have the potential to explore the new anticancer drugs (Table 1).

2.5 *Synthesis of Phosphonyl Spirocyclic Indolines*

Phosphorus-containing compounds are present in all cells of the human body and are involved in a wide range of biological activities in the field of human health and crop protection, such as regular heartbeat and kidney maintenance. In particular, aminophosphonic acid derivatives have proven extremely useful in metabolic regulation, in the enhancement or inhibition studies, and in the development of potential drugs against several metabolic disorders. In this context, introducing phosphorus-containing groups into the spirocyclic indoline skeletons could be a promising strategy to enrich the drug-candidate library.

In 2019, the Vincent group developed a diastereoselective synthesis of α-amino phosphonate derivatives embedded in spirocyclic indolines [24]. The reactions were carried out in the presence of 5.0 equiv. of phosphites as the phosphorus sources and 5.0 equiv. of $Mn(OAc)_3 \cdot 2H_2O$ as the oxidant, and heated to 80°C for 20 h, delivering the corresponding products in acceptable yields. Noteworthily, after extensive screening of the oxidants, only manganese triacetate was competent. Interestingly, the reaction was proposed to be sensitive to the steric hinderance since the di-*tert*-butyl and diphenyl phosphites failed to generate the desired products. Moreover, the corresponding product was only obtained in a 13% yield when diphenyl phosphine oxide was subject to the optimal conditions. Unfortunately, the tolerance of the nucleophilic entity on the C3 side chain was limited to the alcohol motif. Carboxylic acid or tosylamide derivatives proved incompatible in this case (Scheme 19).

To improve the atom economy and decrease the chemical waste and the difficulty of workup, our group devised an electrochemical synthesis of the phosphine-oxide-containing spirocyclic indolines mediated by Cp_2Fe in 2023 [25]. The reactions were

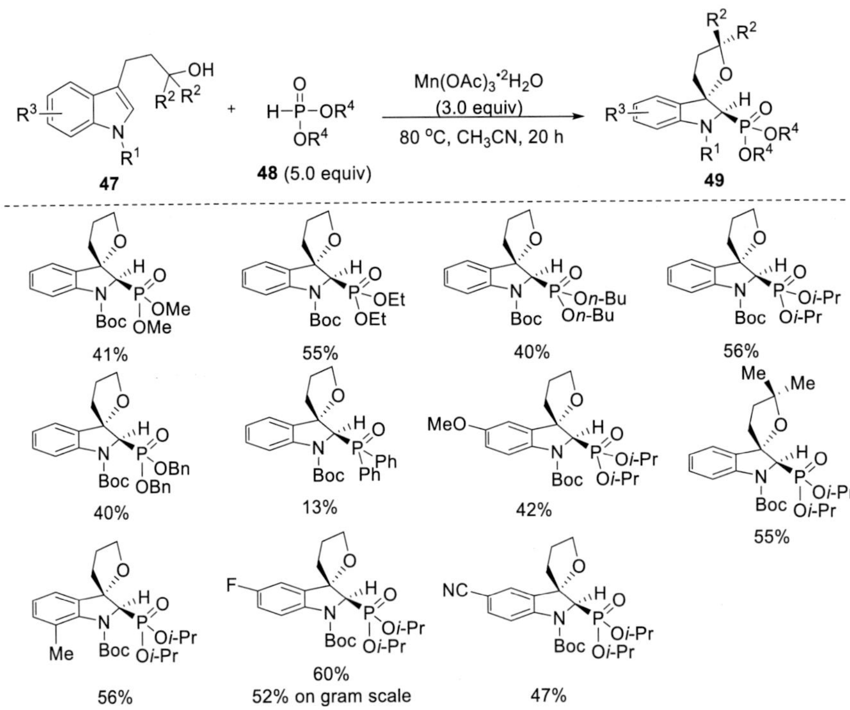

Scheme 19 $Mn(OAc)_3 \cdot 2H_2O$ promoted phosphorylation of indoles

carried out in the undivided cell in the presence of 2.0 equiv. of diphenyl phosphine oxide and 0.2 equiv. of Cp_2Fe, leading to a series of phosphorus-containing products in moderate to good yields. The yield of the dearomatized product **52** could be increased to 74% when the same indole substrate was treated with the electrochemical conditions. This result is notable because the formation of **52** has been proven challenging using either stoichiometric $Mn(OAc)_3{\cdot}2H_2O$ or other oxidants such as $K_2S_2O_8$, CAN, or $Ph_2I^+OTf^-$ were used. This demonstrated that electrochemistry can offer reactivities and selectivities that are not within the purview of most chemical reagents. Moreover, besides pendant alcohols, carboxylic acids were also able to attack the key carbocation intermediate to close the new ring. However, the major product was rearomatized without trapping by the pendant group when a tosylamide was used as the nucleophile. Increasing hindrance or poor nucleophilicity of the tosylamide group could account for the lower yield of the desired indoline. Interestingly, diethyl phosphonate was incompatible to generate the desired product, which could result from the large gap of oxidation potential between diethyl phosphonate and Cp_2Fe (Scheme 20).

From a mechanistic standpoint, ferrocene can be converted to ferrocenium at surface of the anode, then the phosphorus-containing anion **53** could be transformed into *P*-centered radical **54** through a single-electron-transfer process facilitated by ferrocenium in the presence of a base. Subsequently, a radical addition to the indole nucleus **50**, followed by further oxidation, leads to the formation of the crucial carbocation species **56**. Ultimately, the spirocyclic product **52** is afforded through a nucleophilic attack from the side chain onto the carbocation (Scheme 21).

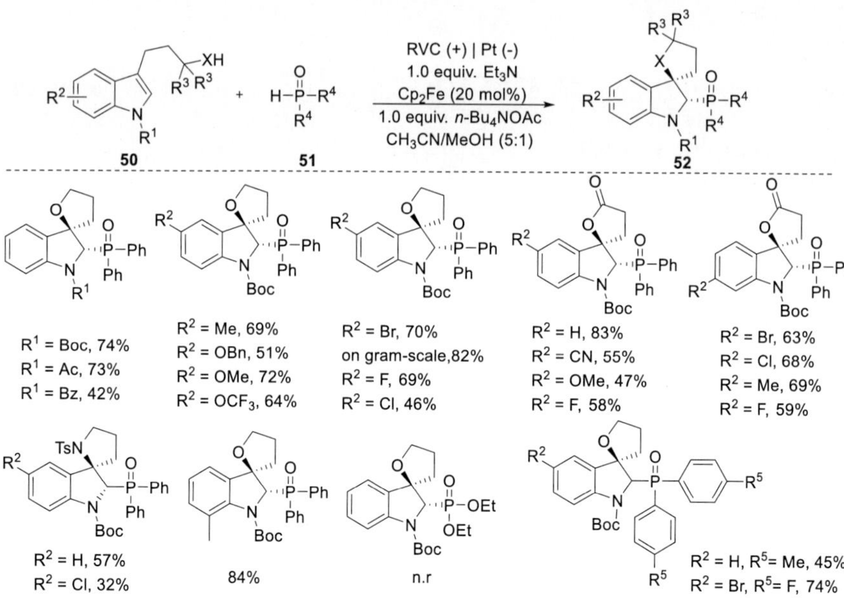

Scheme 20 Cp_2Fe-mediated electrochemical addition of phosphine oxides to indoles

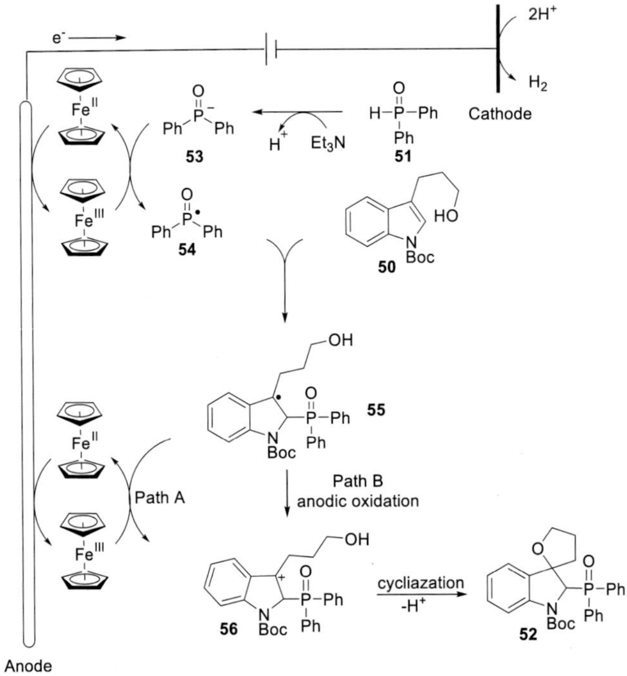

Scheme 21 Mechanistic proposal for Cp_2Fe-mediated electrochemical phosphorylation of indoles

2.6 Conclusion

Electrosynthesis has enjoyed a renaissance in the past decade, which has brought significant growth and attention toward the dearomatization of indoles. Some advancements were facilitated by electrochemistry's ability to harness traceless electrons as oxidants to promote a range of single electron oxidation events, avoiding the usage of stoichiometric oxidants. Moreover, electrocatalysis has been recognized as a viable pathway to generate highly reactive radical intermediates and tune their downstream reactivities to form the functionalized spirocyclic indolines in a single step. To date, electrosynthesis has been applied to the synthesis of methoxylated, hydroxylated, azido, fluorine-containing, or phosphonyl spirocyclic indolines, which can realize the introduction of new C–O, C–N, C–C, or C–P bonds into the spirocyclic indoline frameworks. Remarkably, some trifluoromethylated spirocyclic indolines have already shown anticancer activity in bioactivity studies, implying that functionalized spirocyclic indolines have the potential to be future drug compounds.

Even with these advancements mentioned above, there are still some challenges encountered in the electrosynthesis of spirocyclic indolines. From the perspective of this review, we would like to list them as follow:

1. Development of novel electrochemical activation of indole nucleus to achieve the new reactivity
2. Introduction of other functional groups into the spirocyclic indolines to enrich the compound library
3. Design of new structures of indole substrates to construct different spirocyclic indolines
4. Development of enantioselective syntheses of functionalized spirocyclic indolines
5. Bioactivity studies of spirocyclic indolines to discover the novel drug-like compound

In conclusion, electrochemical dearomatization of indoles is a green, useful, and environmentally benign strategy to enrich the diversity of functionalized spirocyclic indolines and provides some new insights into the synthesis of biologically active indolines, which occupy prominent positions in drug discovery.

Acknowledgments We are grateful for financial support provided by the National Natural Science Foundation of China (Grant NO. 22201150), Natural Science Foundation of Zhejiang Province (No. LQ22B020009), Scientific Research Grant of Ningbo University (215-432000282), Ningbo Top Talent Project (215-432094250), and Ningbo University.

References

1. Molvi KI, Haque N, Awen BZS, Zameeruddin M (2014) Synthesis of spiro compounds as medicinal agents; new opportunities for drug design and discovery. World J Pharm Sci 3:536–563
2. Hiesinger K, Dar'in D, Proschak E, Krasavin M (2020) Spirocyclic scaffolds in medicinal chemistry. J Med Chem 64:150–183
3. Acosta-Quiroga K, Rojas-Pena C, Nerio LS, Gutiérrez M, Polo-Cuadrado E (2021) Spirocyclic derivatives as antioxidants: a review. RSC Adv 11:21926–21954
4. Cihan-Üstündağ L, Şatana D, Erköse-Genç G, Mataracı-Kara E, Çapan G (2019) Design, synthesis, Antitubercular and antiviral properties of new spirocyclic indole derivatives. Monatsh Chem 150:1533–1544
5. Bariwal J, Voskressensky LG, Van der Eycken EV (2018) Recent advances in Spirocyclization of indole derivatives. Chem Soc Rev 47:3831–3848
6. Saraswat P, Jeyabalan G, Hassan MZ, Rahman MU, Nyola NK (2016) Review of synthesis and various biological activities of Spiro heterocyclic compounds comprising oxindole and pyrrolidine moities. Synth Commun 46:1643–1664
7. Hassan AY, Abou-Amra ES (2023) Novel Indoline Spiro derivatives as anticancer agents; design, synthesis, molecular docking studies, and ADME evaluation. Polycycl Aromat Compd 43:6977–6997
8. Wang Q, Song H, Wang Q (2022) Studies on the biological activity of gem-difluorinated 3,3′-spirocyclic indole derivatives. Chin Chem Lett 33:859–862
9. Faraday M (1834) Siebente Reihe von Experimental-Untersuchungen über Elektricität. Ann Phys 109:438
10. Kolbe H (1845) Beiträge zur Kenntniss der gepaarten Verbindungen. Justus Liebigs Ann Chem 54:145–188

11. Frontana-Uribe BA, Little RD, Ibanez JG, Palma A, Vasquez-Medrano R (2010) Organic electrosynthesis: a promising green methodology in organic chemistry. Green Chem 12: 2099–2119
12. Putter H, Hannebaum H (2000) U.S. Patent and Trademark Office, Washington, DC, U.S. Patent No. 6063256
13. Francke R, Little RD (2014) Redox catalysis in organic electrosynthesis: basic principles and recent developments. Chem Soc Rev 43:2492–2521
14. Wu J, Dou Y, Guillot R, Kouklovsky C, Vincent G (2019) Electrochemical dearomative 2, 3-difunctionalization of indoles. J Am Chem Soc 141:2832–2837
15. Wu J, Guillot R, Kouklovsky C, Vincent G (2020) Electrochemical dearomative dihydroxylation and hydroxycyclization of indoles. Adv Synth Catal 362:1712–1719
16. Wu J, Abou-Hamdan H, Guillot R, Kouklovsky C, Vincent G (2020) Electrochemical synthesis of 3a-bromofuranoindolines and 3a-bromopyrroloindolines mediated by $MgBr_2$. Chem Commun 56:1713–1716
17. Li J, Liu M, Li Q, Tian H, Shi Y (2014) A facile approach to spirocyclic 2-azido indolines via azidation of indoles with ceric ammonium nitrate. Org Biomol Chem 12:9769–9772
18. Mo K, Zhou X, Wu J, Zhao Y (2022) Manganese-mediated electrochemical dearomatization of indoles to access 2-azido spirocyclic indolines. J Org Chem 87:16106–16110
19. Ryzhakov D, Jarret M, Guillot R, Kouklovsky C, Vincent G (2017) Radical-mediated dearomatization of indoles with sulfinate reagents for the synthesis of fluorinated spirocyclic indolines. Org Lett 19:6336–6339
20. Ye JH, Zhu L, Yan SS, Miao M, Zhang XC, Zhou WJ, Li J, Lan Y, Yu DG (2017) Radical trifluoromethylative dearomatization of indoles and furans with CO_2. ACS Catal 7:8324–8330
21. Mei L, Moutet J, Stull SM, Gianetti TL (2021) Synthesis of CF_3-containing spirocyclic indolines via a red-light-mediated trifluoromethylation/dearomatization cascade. J Org Chem 86:10640–10653
22. Mo K, Zhou X, Wu J, Zhao Y (2022) Electrochemical dearomatization of indoles: access to diversified fluorine-containing spirocyclic indolines. Org Lett 24:2788–2792
23. Lan J, Li S, Lin K, Zhou P, Chen W, Gao L, Zhu T (2022) The eco-friendly electrosynthesis of trifluoromethylated spirocyclic indolines and their anticancer activity. Org Biomol Chem 20: 3475–3479
24. Ryzhakov D, Jarret M, Baltaze JP, Guillot R, Kouklovsky C, Vincent G (2019) Synthesis of 3, 3-spirocyclic 2-phosphonoindolines via a dearomative addition of phosphonyl radicals to indoles. Org Lett 21:4986–4990
25. Mo K, Zhou X, Wang J, Wu J, Zhao Y (2023) Cp_2Fe mediated electrochemical synthesis of phosphonylated spirocyclic indolines via dearomatization of indoles. Org Lett 25:3956–3960

Top Heterocycl Chem (2024) 61: 113–134
https://doi.org/10.1007/7081_2024_77

Published online: 9 August 2024

Construction of *N*-Heterocycles via Electrochemically Generated Nitrogen Radicals and Nitrilium Ions

Zhaojiang Shi and Ke-Yin Ye

Contents

Abstract Nitrogen-containing heterocycles widely exist in natural products, medicinal compounds, and functional materials. The recent renaissance of organic synthetic electrochemistry provides many novel approaches to *N*-heterocycles. During the last several years, our group has been devoted to developing green and sustainable electrochemical methodologies to synthesize diverse *N*-heterocyclic scaffolds that are otherwise difficult or impossible to achieve. The mechanistic investigations allow us to take the advantage of electrochemical proton-coupled electron-transfer (PCET) strategy and electrochemically in-situ-generated Ritter-type intermediates for rapid access toward a wide range of value-added *N*-heterocycles, including benzoxathiazine dioxides, sultam-fused pyridinones, isoindolinones, tetrazoles, and others.

Keywords Electrochemistry · *N*-heterocycle · Nitrogen radical · Proton-coupled electron transfer · Ritter reaction

Z. Shi and K.-Y. Ye (✉)
Key Laboratory of Molecule Synthesis and Function Discovery (Fujian Province University), College of Chemistry, Fuzhou University, Fuzhou, China
e-mail: kyye@fzu.edu.cn

1 General

N-heterocycles are widely found in pharmaceuticals, agrochemicals, and synthetic materials (Fig. 1) [1–4]. The introduction of nitrogen atoms in pharmaceuticals has a great influence on their pharmacokinetic profile [5], originating from the fine-tuning of physical and chemical properties [6]. Free-radical chemistry has become a valuable tool for constructing diverse *N*-heterocycles [7]. Compared with conventional polar bond-forming approaches, the innate reactivity of a radical strategy provides new disconnection logic and potentially minimizes the interconversion of functional groups and the installation and removal of protecting groups. Among them, nitrogen-centered radicals (NCRs) are reactive radical intermediates that have found broad applications in organic synthesis to construct *N*-heterocycles.

In recent years, organic synthetic electrochemistry has been experiencing dynamic development. As a green and sustainable tool, electrosynthesis also serves as a robust and enabling handle for exploring new reactivity. Indeed, a vast number of nitrogen-containing heterocycles have been synthesized through various electrochemical strategies [8, 9]. Among them, we are particularly interested in the adoption of the electrochemical proton-coupled electron-transfer (PCET) process and electrochemically generated nitrilium ions to construct *N*-heterocycles (Fig. 2).

2 Electrochemical Synthesis of *N*-Heterocycles via PCET-Generated Nitrogen Radicals

Proton-coupled electron transfer is a redox process in which electrons and protons are transferred together and can be used as a very powerful tool for breaking strong and inert chemical bonds to generate highly reactive free-radical species

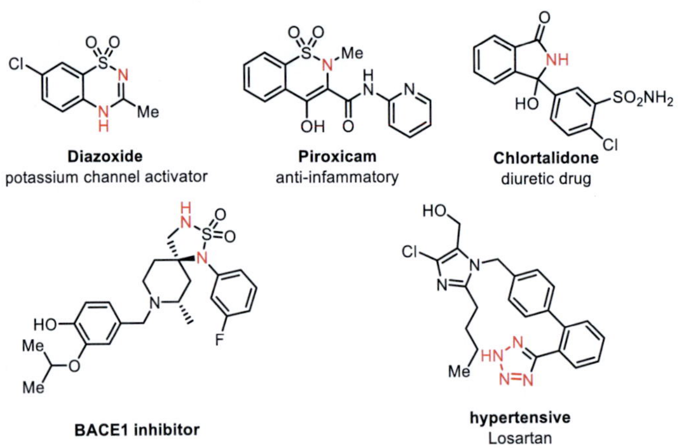

Fig. 1 Bioactive molecules with *N*-heterocycles

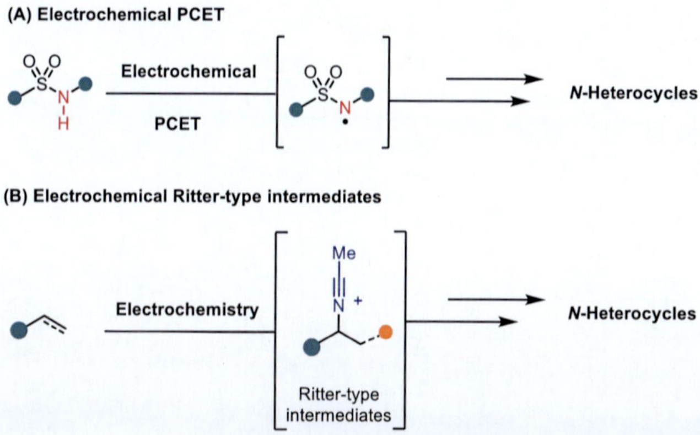

Fig. 2 *N*-Heterocycles via electrochemically generated nitrogen radicals and nitrilium ions

Fig. 3 Electrochemical PCET for the generation of nitrogen radicals

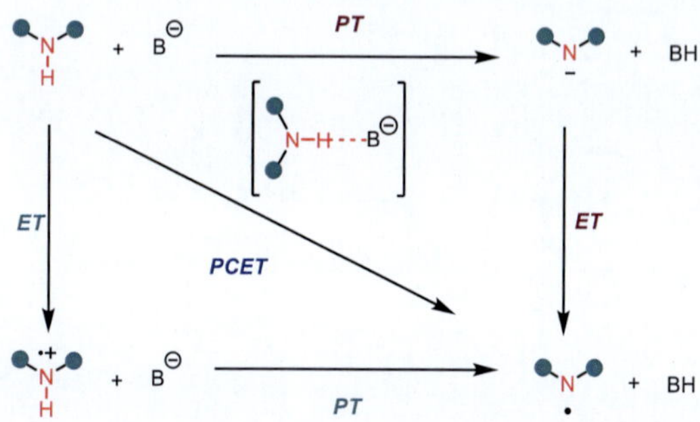

[10–12]. Among various PCET-facilitated generations of free radicals [13, 14], nitrogen radical species can also be readily formed through an electrochemical PCET strategy [15] that further enables the construction of diverse *N*-heterocycles (Fig. 3).

2.1 *Electrochemical Migratory Cyclization and* Ortho-*Cyclizations*

The ubiquity of PCET processes in biological redox systems [13, 16] inspired us to propose an electrochemical PCET process that might be applicable for the facile preparation of sulfonamidyl free radicals. Following this hypothesis, we first chose a simple and easily available *N*-acylsulfonamide (**1**) as the starting material under electrochemical reaction conditions to realize an intramolecular *N*-radical migratory cyclization [17]. The extensive investigation of reaction parameters shows that the intramolecular radical migration cyclization of *N*-acylsulfonamides (**1**) can be effectively realized by constant-current electrolysis (CCE = 15 mA, carbon cloth anode, and Pt cathode) at room temperature to synthesize benzoxathiazine dioxides

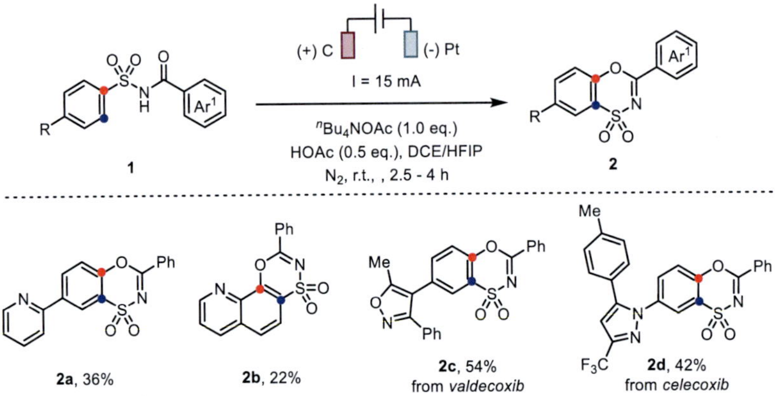

Scheme 1 Electrochemical migratory cyclization of *N*-acylsulfonamides

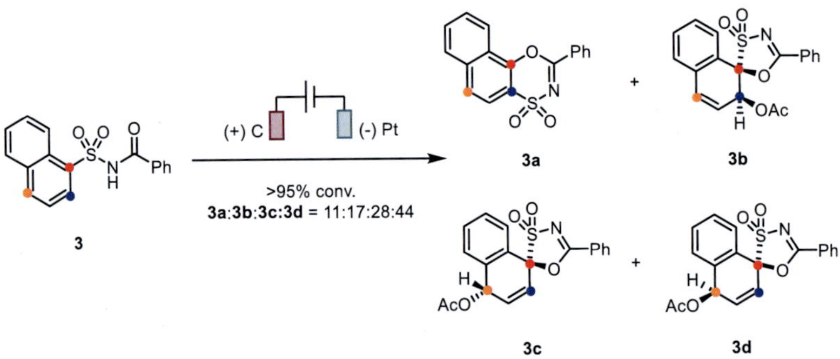

Scheme 2 Electrolysis of *N*-(naphthalen-1-ylsulfonyl)benzamide (**3**)

(**2**, Scheme 1). Under the optimized conditions, it was found that heterocyclic substituents (**2a-2b**) and drug derivatives (**2c-2d**) could undergo the anticipated migratory cyclization reaction to obtain the corresponding biologically important benzoxathiazine dioxide derivatives [18].

Notably, the electrolysis of naphthalene derivative (**3**) of lower aromaticity afforded not only the expected migratory cyclization product (**3a**) but also several unexpected dearomatized spirocyclic products (**3b**–**3d**, Scheme 2). For the dearomative spirocyclization, the nucleophilic addition of an acetate ion occurred either at the C2 (**3b**) or the C4 (**3c** and **3d**) positions of the naphthalene scaffold. The adjacent C2-functionalized spiro ring (**3b**) was obtained as a single diastereomer, while the selectivities of the C4-functionalized spirocycles were low. These findings demonstrated that the above-mentioned migratory cyclization may involve a spirocyclic intermediate.

Nuclear-magnetic-resonance (NMR) studies indicated that an intermolecular hydrogen bond exists between the N – H group of **1a** and the acetate anion (Fig. 4). Consistent with this observation, cyclic voltammetry studies disclosed

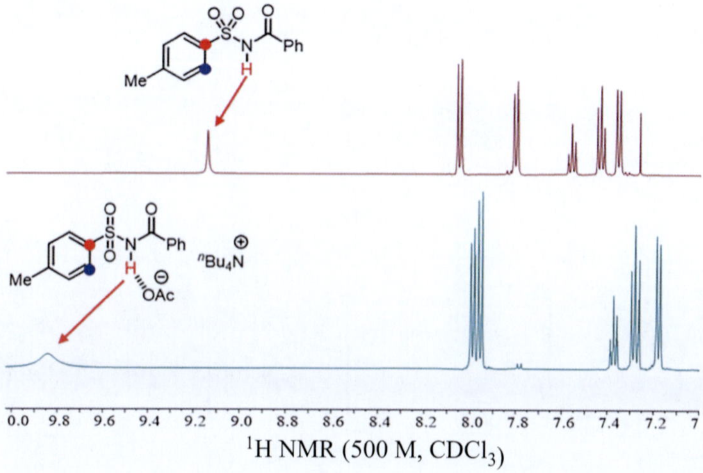

Fig. 4 NMR titration experiments

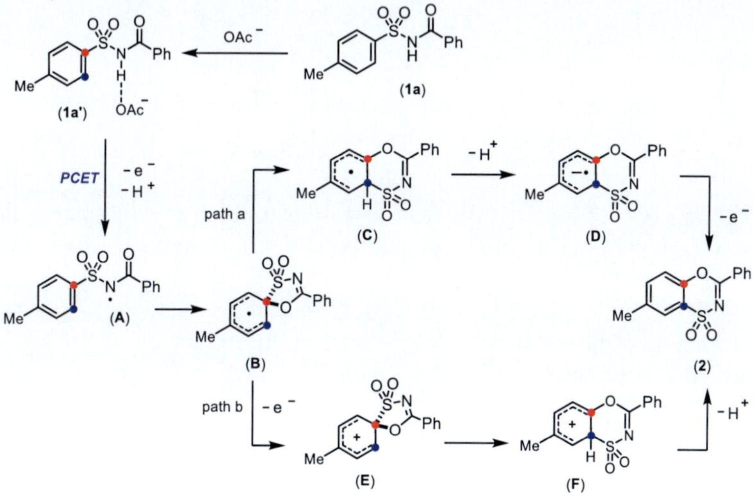

Scheme 3 Proposed mechanism for the electrochemical migratory cyclization

that the oxidation potential of *N*-tosylbenzamide (**1a**, $E_{ox} = 2.40$ V) was much higher than that of $^nBu_4NOAc$ ($E_{ox} = 0.92$ V). However, adding an equal molar ratio of $^nBu_4NOAc$ to the *N*-tosylbenzamide solution resulted in the decrease of its oxidation current, and concurrently, a new oxidation peak emerged (1.18 V).

Taking the results of cyclic voltammogram and NMR titration experiments together, an electrochemical PCET process of acid-base complex (**1a′**) from the N – H group in **1a** and acetate anion for the facile generation of an *N*-centered radical species (**A**) was proposed (Scheme 3). The in-situ-formed intermediate (**A**) then generates spirocyclic cyclohexadienyl radical (**B**). We anticipate that the first

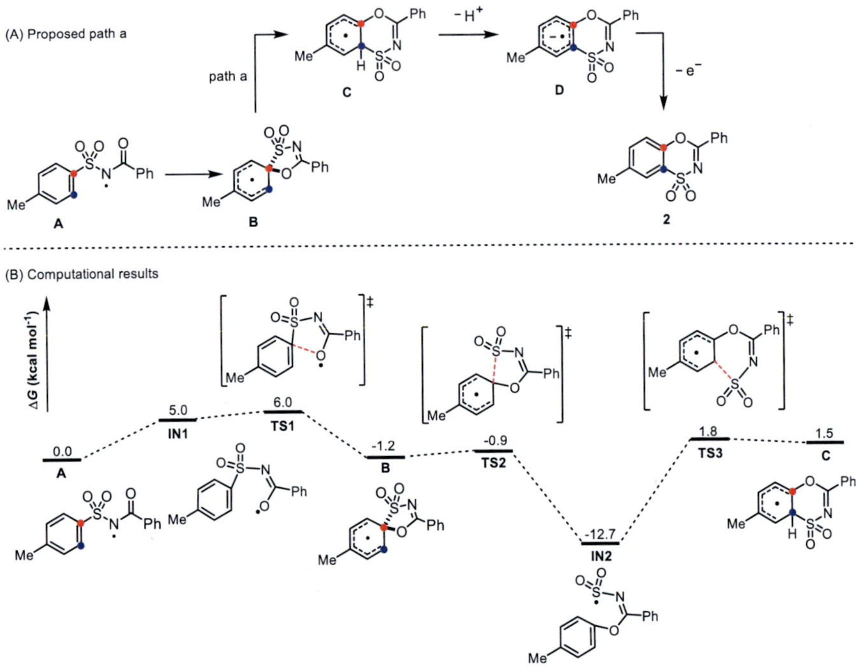

Scheme 4 The potential energy surface of the path a

possible reaction pathway is the radical rearomatization process (path a), which includes a radical-type sulfone migration (**B** to **C**) and subsequent base-promoted homolytic aromatic substitution (**D**). Alternatively, the spirocyclic cyclohexadienyl radical (**B**) might undergo a further single-electron oxidation to generate the corresponding carbocation (**E**). A cationic sulfone migration (**E** to **F**) is then followed by base-facilitated deprotonation to deliver the migratory cyclization product (**2**).

We further studied the two above-mentioned reaction pathways by density-functional-theory (DFT) calculations (Scheme 4). According to our calculation results, the *N*-centered radical species **A** is readily converted into its oxygen radical **IN1** with a moderate reaction barrier of 6.0 kcal mol^{-1}, which is further transformed into the spirohexadienyl radical **B**. In path a, intermediate **B** achieves the highly chemoselective S–C bond cleavage via **TS2** with a reaction barrier of only 0.3 kcal mol^{-1}, resulting in the sulfonyl radical **IN2**. A subsequent radical cyclization from **IN2** forms the fused cyclohexadienyl radical **C** via **TS3** (reaction barrier, 14.5 kcal mol^{-1}).

By contrast, path b involves a key spirocyclic cyclohexadienyl carbocation **E** (Scheme 5). The DFT calculations suggest that the sulfone migration of **E** to the adjacent arene moiety is through a transition state **TS4** with a reaction barrier of 3.9 kcal mol^{-1} to obtain a six-membered ring cation intermediate **F**. Interestingly, the following rearomatization processes were found to be favored through the

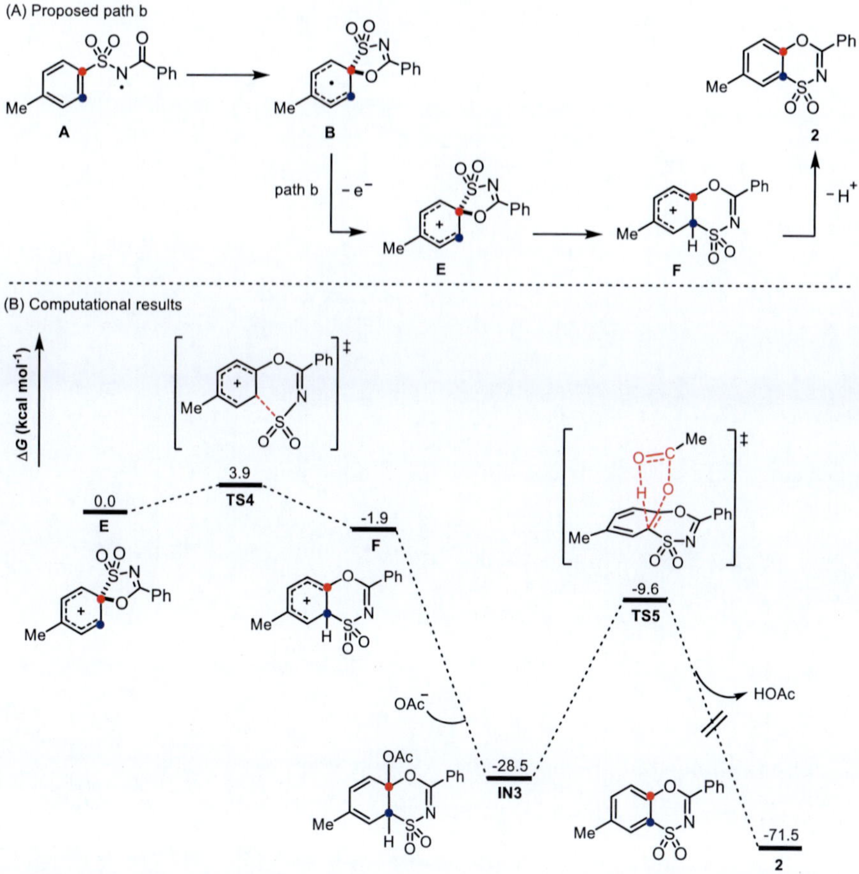

Scheme 5 The potential energy surface of path b

participation of the acetate. Based on the above calculation results, both the radical and cationic sulfone migration pathways are viable. These mechanistic insights pave the theoretical basis for our following research on the electrochemical synthesis of *N*-heterocycles (vide infra).

Radical-mediated cascade reactions are well-documented in the literature [19–22]. However, very little attention has been paid to the migratory attitude of different linkers of radical intermediates, such as in the in-situ-generated spirocyclohexadienyl radicals. It is fundamentally interesting and also potentially informative for the design of selective radical transformations via the in-depth understanding of the innate migratory attitude of spirocyclic intermediates. In this regard, we recently developed an electrochemical selective migration cyclization or cascade *ortho*-cyclization of 2-alkynylbenzenesulfonamides (Scheme 6) [23]. Though the radical cascade *ortho*-cyclization of 2-alkynylbenzenesulfonamides can be expected, the migratory cyclization products

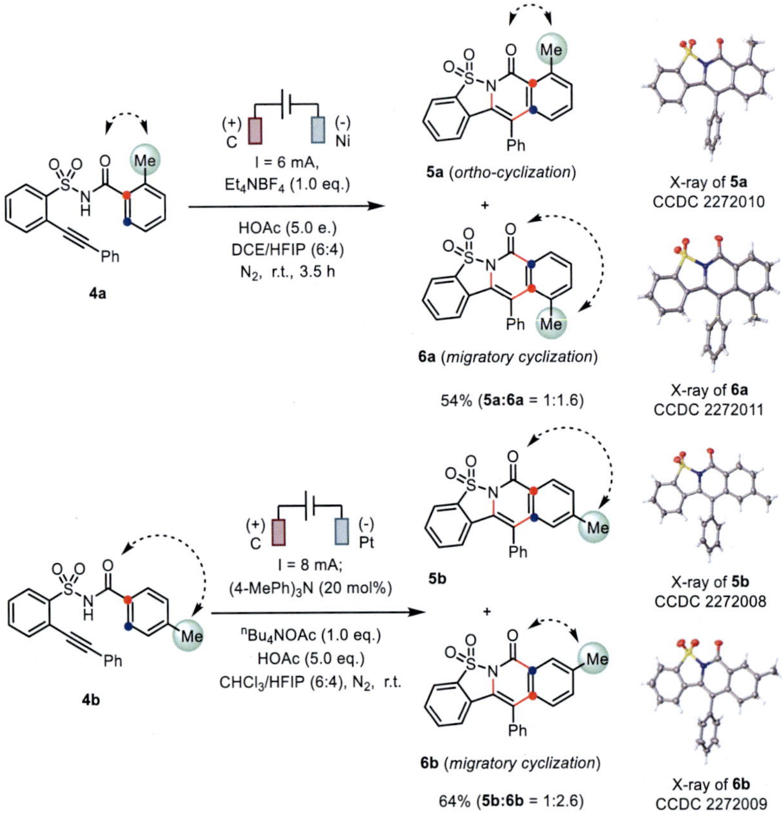

Scheme 6 Electrochemical cascade migratory versus *ortho*-cyclization of 2-alkynylbenzenesulfonamides

through spirocyclic intermediates are rarely studied. Notably, the use of *ortho*-methyl substrate (**4a**) resulted in both the cascade *ortho*-cyclization (**5a**) and migratory cyclization (**6a**) products, both of which were isolated and characterized by the X-ray diffraction analysis. While the location of *ortho*-methyl remained the same for **5a**, the migratory product (**6a**) featured a translocation of the original *ortho*-methyl of the benzoate moiety to the *meta*-position. A similar situation was encountered with *para*-methyl substrate (**4b**). This current strategy provides rich diversity for the synthesis of sultam-fused pyridinones, which exhibit promising antibacterial activity [24].

A possible mechanistic rationale is illustrated in Scheme 7. Firstly, the N–H bond of the substrate (**4a**) forms a hydrogen bond with acetate as an acid-base complex, which undergoes an electrochemical PCET process to deliver the nitrogen radical (**4a′**). This radical species can undergo intramolecular cyclization to form the vinyl radical (**A**) with two possible follow-up reaction pathways. The first pathway is its attack into the *ortho* position of the benzene ring, followed by electrochemical

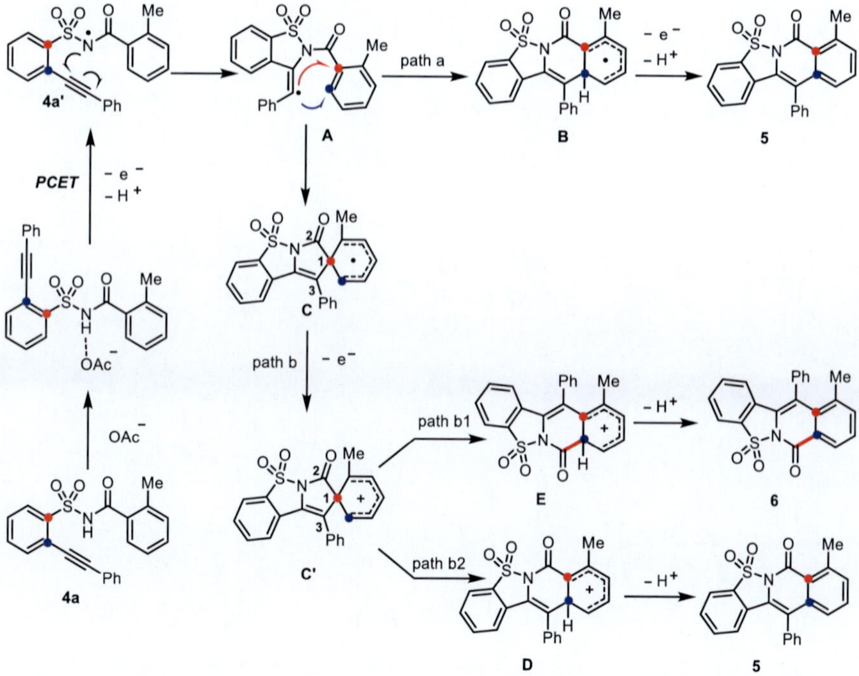

Scheme 7 Mechanistic rationale for the *ortho*-cyclization and migratory cyclization

single-electron oxidation and deprotonation to give rise to the *ortho*-cyclization product (path a). In path b, however, an *ipso*-addition of the nitrogen radical (**A**) generates the spirocyclic radical intermediate (**C**), which is further electrochemically oxidized to its cationic species (**C′**). At this stage, selective vinyl- and acyl-migrations should provide the *ortho*-cyclization (**5**) and migratory-cyclization (**6**) products, respectively.

We further studied the migratory selectivity by DFT calculations (Scheme 8). It is found that both radical addition pathways for the 2-methyl-substituted vinyl radical (**A**) were thermodynamically and kinetically favorable, where the reaction barrier of *ipso*-addition is only 1.8 $kcal{\cdot}mol^{-1}$ lower than that of *ortho*-addition (**TS1** versus **TS1′** is 10.7 versus 8.9 kcal mol^{-1}).

Since the spirocyclic radical (**C**) is readily oxidized into its corresponding cation intermediate (**C′**) via a radical-polar-crossover process, we decided to investigate the possibility of the spirocyclic cation (**C′**) as the key intermediate in migratory cyclization. Indeed, the spirocyclic cation (**C′**) displayed an elongated C_{ipso}–C_1 bond (1.662 Å versus 1.569 Å) and a shortened C_{ipso}–C_2 bond (1.536 Å versus 1.556 Å) than those of the radical intermediate (**C**). These observations are consistent with the hypothesis that the acyl-migration of the spirocyclic cation intermediate should be more accessible than the vinyl migration (Scheme 9).

Starting from the spirocyclic cation (**C′**), the acyl-migration is found to be more favorable than vinyl migration both kinetically and thermodynamically (Scheme 10).

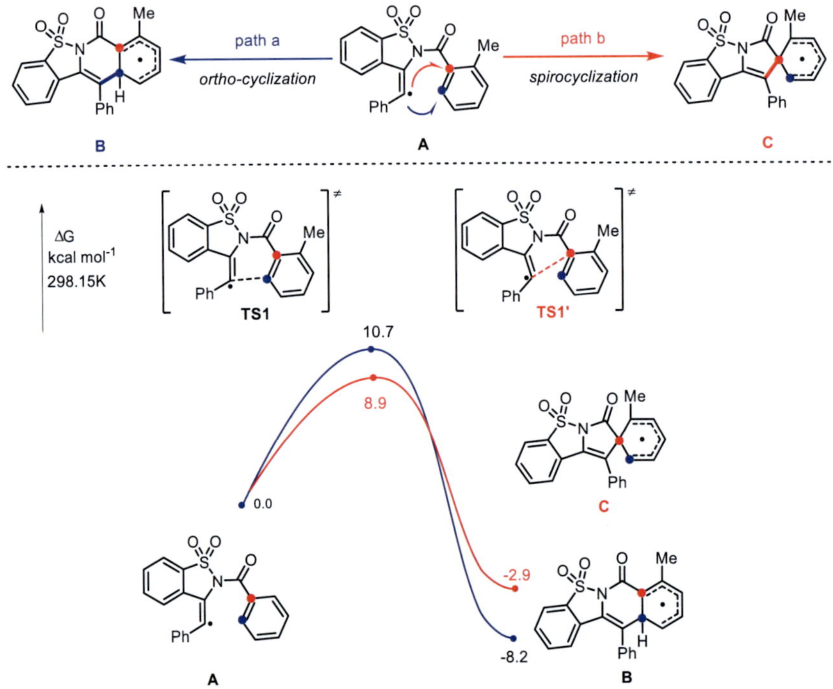

Scheme 8 Radical addition pathways for the 2-methyl-substituted vinyl radical (**A**)

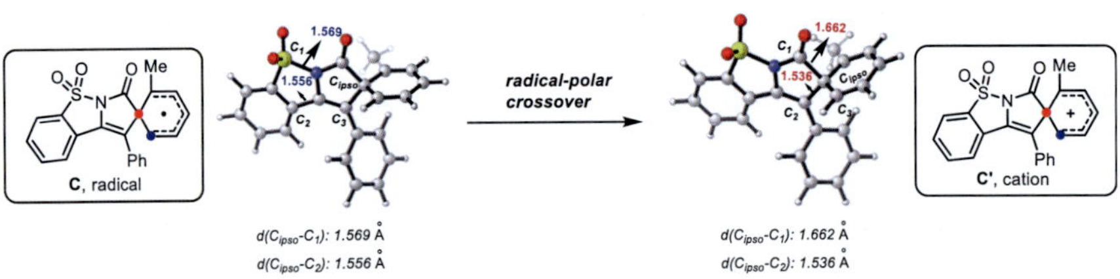

Scheme 9 Bond lengths in vinyl radical and vinyl cation

However, the small difference in reaction barriers between these two transition states (7.6 vs. 8.1 kcal mol^{-1}) suggests that the migratory cyclization (**E**) is only modestly selective, which is also consistent with our experimental observation (ratios of migratory cyclization versus *ortho*-cyclization up to 8:1).

Interestingly, when the alkynyl moiety was installed at the benzoyl side (**7**), only the electrochemical intramolecular 5-*exo-dig* aza-cycloaddition reaction occurred to afford a series of isoindolinone derivatives (Scheme 11) [25]. This reaction was again proposed to be initiated from the nitrogen radical generated via the electrochemical PCET process, followed by *5-exo-dig* cyclization and iterative oxidation/nucleophilic capture by acetate. Besides acetate, propionic acid also functioned as a competent terminating agent (**8d**).

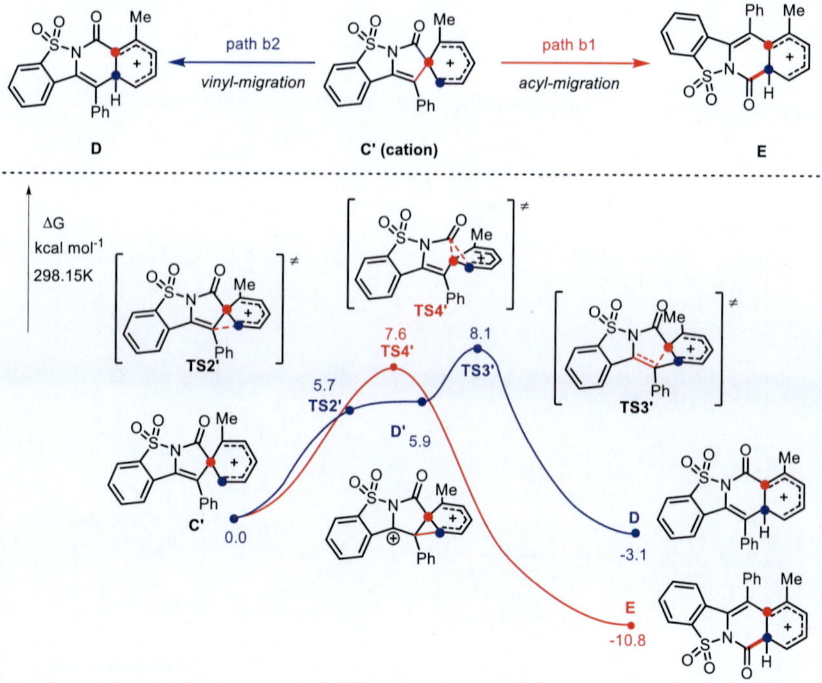

Scheme 10 DFT calculations on vinyl-and acyl-migration of the spirocation

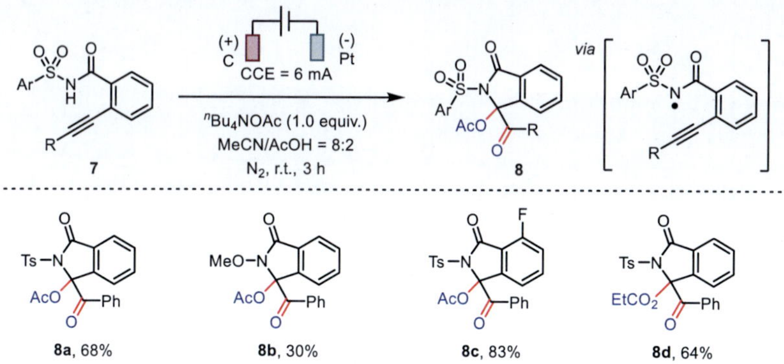

Scheme 11 The electrochemical 5-*exo-dig* aza-cycloaddition of 2-alkynylbenzamides

2.2 Electrochemical Dearomative Spirocyclization

When *N*-acyl thiophene-2-sulfonamide (**9**) was used as the substrate, an intramolecular 5-*exo-trig* dearomative spirocyclization reaction was realized (Scheme 12) [26]. This reaction proceeds via a nitrogen-centered radical and its subsequent intramolecular spirocyclization. The ready oxidation of thiophene moiety then

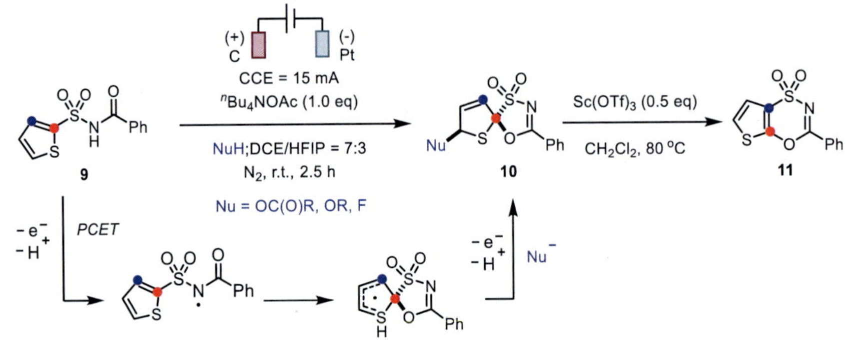

Scheme 12 Electrochemical dearomative spirocyclization of *N*-acyl thiophene-2-sulfonamide

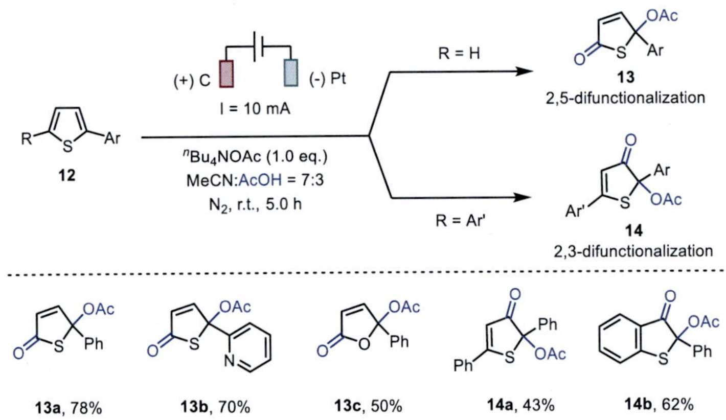

Scheme 13 Electrochemical oxidative dearomatization of 2-arylthiophenes

generates a thiophene cation that is intercepted by a nucleophile to afford the dearomatized heterocycles. Various nucleophiles, including carboxylates, alcohols, and fluoride can be easily employed to access remotely functionalized spirocycles of promising anti-tumor activity. $Sc(OTf)_3$ enabled the rearomatization of dihydrothiophene (**10**) via the selective cleavage of the carbon–sulfur bond over the carbon–oxygen bond.

The electrochemical oxidative dearomatization of thiophenes can also be realized in an intermolecular manner (Scheme 13) [27]. Various 2-arylthiophenes (**12**) underwent a selective electrochemical oxidative dearomatization to afford a series of dearomatized dihydrothiophenones. Remarkably, a wide array of functionalities, such as phenyl (**13a**), pyridine (**13b**), and furan ring (**13c**) were well tolerated. When C2 and C5 positions are both substituted by phenyl groups, the anticipated dearomatization reactivity was still observed but gave the vicinal difunctionalized dihydrothiophenone (**14a**). A similar observation was recorded with benzothiophene derivatives (**14b**).

2.3 Continuous-Flow Electrosynthesis

Compared with electrochemical batch reactors, continuous-flow electrochemical microreactors are featured for their ease of scale-up preparation and short reaction times. The distance between two electrodes in electrochemical continuous-flow microreactors is short and, more importantly, adjustable. In addition, the use of supporting electrolytes can be reduced or even eliminated in some cases [28, 29].

In an electrochemical microreactor, we found that the *N*-acylsulfonamides (**15**) underwent an electrochemical oxidative dearomatization to provide a series of spirocycles (**16**, Scheme 14) [30]. The substrate scope of this reaction is broad and the targeted products were obtained in moderate to good yield. Notably, upon treatment with $FeCl_3$, an exclusive sulfone migration/rearomatization took place to afford the biologically interesting hydroxylated benzothiazine dioxide (**17a**), which is otherwise challenging to prepare.

To verify the practicability of this reaction in scale-up preparation, we carried out a decagram-scale synthesis of spirodienone (**16a**) in 61% yield (10.57 g) by the continuous-flow electrolysis of 63 mmol of *N*-(4-methoxyphenyl)sulfonyl) benzamide at a current of 130 mA (Fig. 5). This result highlights the synthetic potential of continuous-flow electrosynthesis.

Similarly, a series of biologically active spirodienone derivatives (**19**) can also be efficiently and conveniently synthesized by oxidative dearomatization in a flow

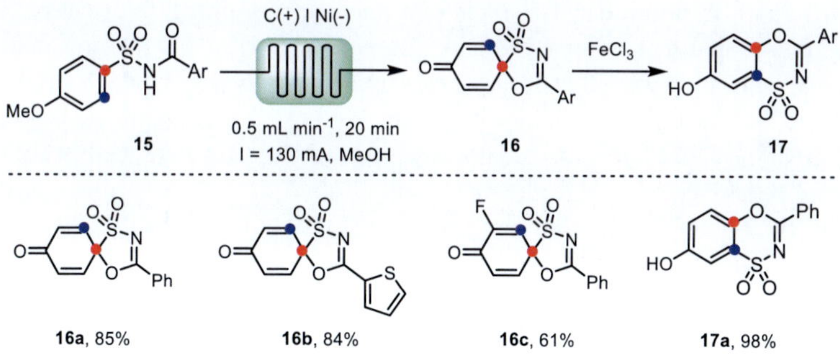

Scheme 14 Electrochemical dearomative spirocyclization of *N*-acyl sulfonamides in a continuous-flow cell

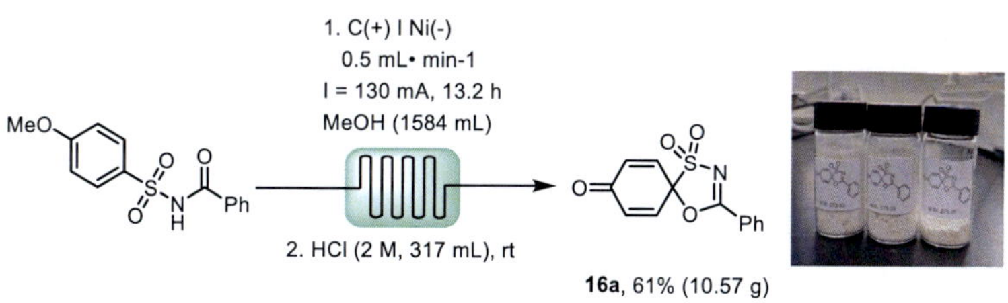

Fig. 5 Decagram-scale synthesis in a continuous-flow electrochemical microreactor

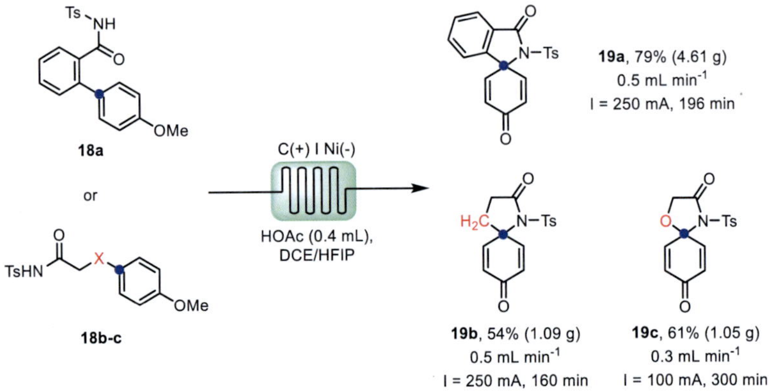

Scheme 15 Electrochemical synthesis of spirodienones in a continuous-flow cell

electrochemical microreactor (Scheme 15) [31, 32]. Besides the biphenyl skeleton (**18a**), the tethers between methoxyphenyl and amide moieties could be all-carbon (**18b**), or oxygenated (**18c**) linear chains. In addition, the incorporation of a continuous-flow microreactor rendered the gram-scale preparation of these analogous compounds readily accessible.

Compared with the achiral sulfonyl fluorides, sulfonimidoyl fluorides are chiral linking agents in the sulfur (VI) fluoride exchange reaction (SuFEx). However, conventional syntheses typically need to use unstable sulfonyl chlorides, toxic and corrosive sulfur fluoride (SOF_4), or expensive electrophilic fluorinating agents. Using the simple and easily available sulfinamide (**20**) as the starting material, a series of sulfonimidoyl fluorides were synthesized through the electrochemical oxidative nucleophilic fluorination (Scheme 16) [33]. By using an electrochemical continuous-flow microreactor, the gram-scale products (**21**) were readily accessible. The rich reactivity of SuFEx chemistry was further proved by various transformations of the obtained sulfonimidoyl fluoride (**21**) to its CF_3-substituted sulfonimide (**21a**), sulfonimide amide (**21b-c**), and sulfonimide ester derivatives (**21d-e**).

Regarding the mechanism (Scheme 16B), an acid-base complex (**B**) from the sulfinamide (**A**) and base proceeds with PCET to generate a sulfinamidyl radical (**C**), followed by successive electron oxidation to give a sulfonimidoyl cation (**D**), which is readily intercepted by the nucleophilic fluoride. DFT calculations suggest the sulfonimidoyl cation exists as a trigonal-planar geometry ($\Sigma S^{CON} = 359°$) and the use of the optically pure chiral sulfinamide did not lead to the enantioenriched sulfonimidoyl fluorides under the current electrolytic conditions.

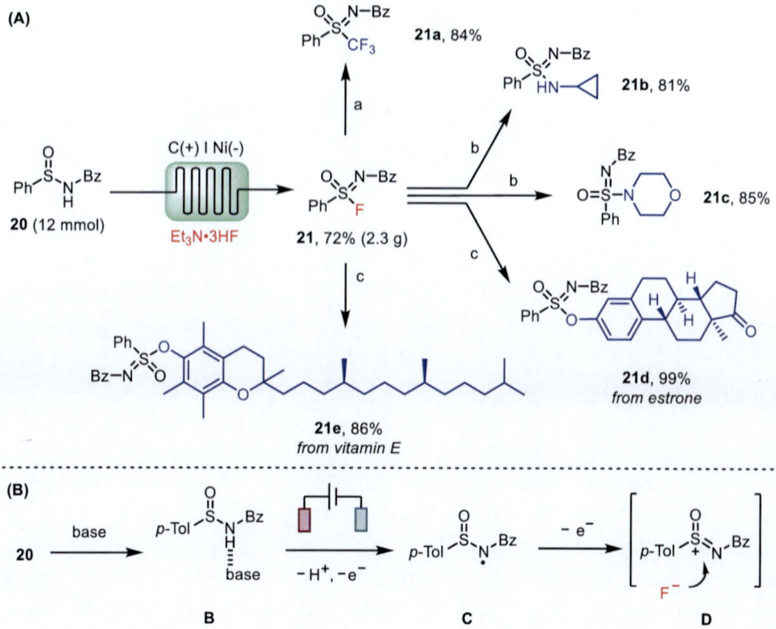

Scheme 16 Continuous-flow electrosynthesis of sulfonimidoyl fluoride and SuFEx linkage reactivity. For the flow electrolysis, Q = 22.8 F mol^{-1}, t_R = 2.5 min. For the derivatizations, reaction conditions: (**a**) $TMSCF_3$ (2.0 equiv.), $^{n}Bu_4NF$ (0.5 equiv.), THF, 0°C to r.t., 12 h; (**b**) amine (2.0 equiv.), Et_3N (2.0 equiv.), LiBr (2.0 equiv.), MeCN, 80°C, 24 h (cyclopropylamine for **21b**, morpholine for **21c**); (**c**) phenol derivatives (1.0 equiv.), DBU (1.0 equiv.), MeCN, r.t., 30 min (estrone for **21d**, vitamin E for **21e**)

3 Electrochemical Synthesis of *N*-Heterocycles via In-Situ-Generated Ritter Intermediates

The use of cheap and abundant nitriles and olefins to synthesize value-added amides bears great advantages and broad application prospects. As early as 1948, Ritter and Minieri found that nitriles readily react with olefins or tertiary alcohols under acidic conditions to form amides (Scheme 17a). When nitriles are replaced by HCN, the obtained formamides are further hydrolyzed to afford alkylamines [34]. Mechanistic investigations revealed that this reactivity proceeds via a key nitrilium ion, i.e., Ritter intermediate, that is generated from the nucleophilic addition of the nitrile to the carbenium ion. This reaction, now termed the Ritter reaction, has found broad applications in organic synthesis (Scheme 17b).

Tetrazoles are bioisosteres for carboxylic acids, furan, and amide moieties, as well as having other beneficial physicochemical properties [35, 36], which renders them an important pharmacophore in medicinal chemistry. In addition, the nitrogen-rich character of tetrazoles also facilitates their use as a popular explosophore in high-energy materials [37, 38]. Therefore, the rapid development of pharmaceutical,

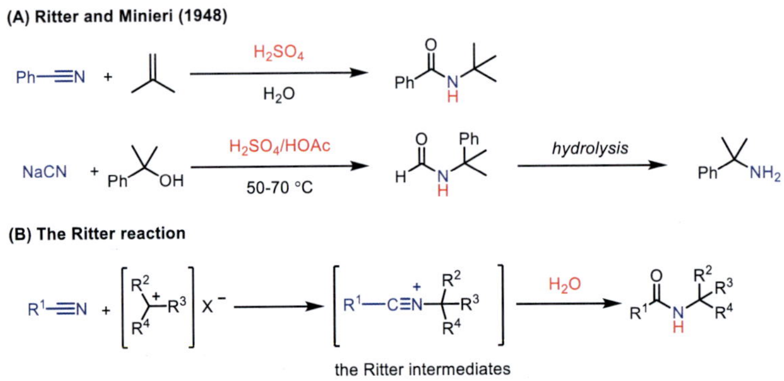

Scheme 17 The Ritter reaction

biology, and materials sciences depends on efficient, straightforward, and modular synthetic approaches to diverse and novel tetrazole scaffolds.

Recently, electrochemistry has been widely used in direct C(sp^3)–H functionalizations [39] owing to its robustness in organic synthesis and unique capability in controlling reactivity and selectivity via applied potentials [40, 41]. Therefore, we took advantage of an electrochemical multicomponent reaction (e-MCR) in the direct C(sp^3)–H tetrazolation (**22**) by using saturated hydrocarbons, nitrile, and azide as the starting materials (Scheme 18) [42]. Notably, we also found that the synthetic strategy can be further transferred into the electrochemical vicinal azido tetrazolation of alkenes. The resulting vicinal azidotetrazoles (**23**) were further demonstrated to be versatile building blocks in diverse synthetic elaborations. Both the electrochemical C(sp^3)–H tetrazolation and vicinal azido tetrazolation reactions display a broad substrate scope and are also compatible with various functional groups and drug derivatives (**22a-23b**).

Azido and tetrazole moieties are frequently spread as the explosophores in high-energy materials. Therefore, differential scanning calorimetry (DSC) and thermogravimetric (TG) analyses were employed to probe the thermal sensitivity of those vicinal azidotetrazoles (Fig. 6). Indeed, the high decomposition enthalpy (−1,270 J g^{-1} or −291 kJ mol^{-1}) of the azidotetrazole (**23c**) implies that its scale-up preparation and usage would need special safety precautions. In contrast, its high onset temperature (233°C) reflects a relatively insensitivity to heat. This hypothesis is further substantiated by its TD_{24} data as a consecutive prediction of a reasonable safety margin. Finally, the thermal decomposition pathway of this azidotetrazole is proposed to initiate via the extrusion of the tetrazole moiety.

Based on our mechanistic investigations, we proposed a possible reaction mechanism for the electrochemical C(sp^3)–H tetrazolation (Scheme 19a). Firstly, ethylbenzene (**A**) undergoes a single-electron oxidation to generate a phenyl radical-cation intermediate, followed by a deprotonation to deliver a benzyl radical (**B**), which then readily undergoes a second single-electron oxidation at the anode to obtain a benzyl carbocation (**C**). Then, the nucleophilic attack of **C** by MeCN forms

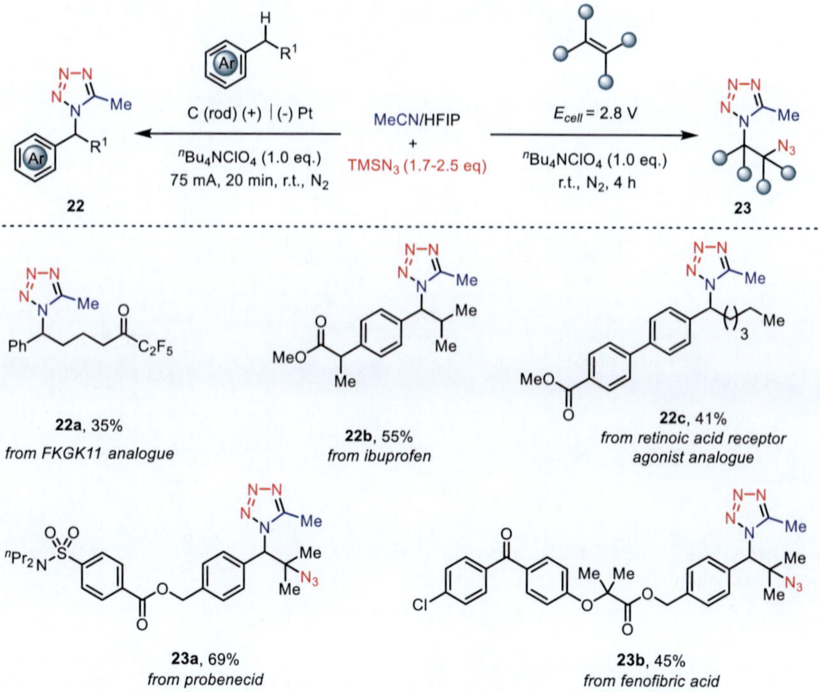

Scheme 18 Electrochemical C(sp^3)–H tetrazolation and vicinal azidotetrazolation

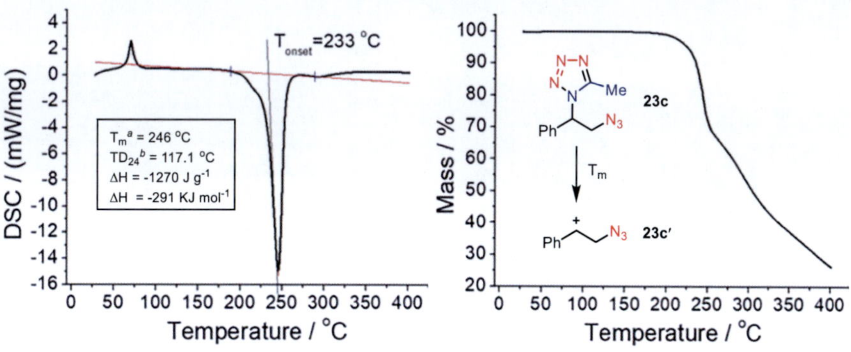

Fig. 6 DSC (left) and TG (right) curves of the azidotetrazole (**23c**). [a] Temperature of maximum in the DSC curve; [b] Temperature in which the time of maximum rate is 24 h

a Ritter-type intermediate (**D**), which finally reacts with azide to afford C–H tetrazolation product (**22**).

Similarly to the C(sp^3)–H tetrazolation of alkyl arenes, the β-azido carbocation (**II**), generated via the sequential addition of the azido radical ($N_3^{\bullet}$) to an alkene and the single-electron oxidation or the resulting benzylic radical (**I**), is the key intermediate in the azidotetrazolation reaction (**III**, Scheme 19b). Notably, the potentially

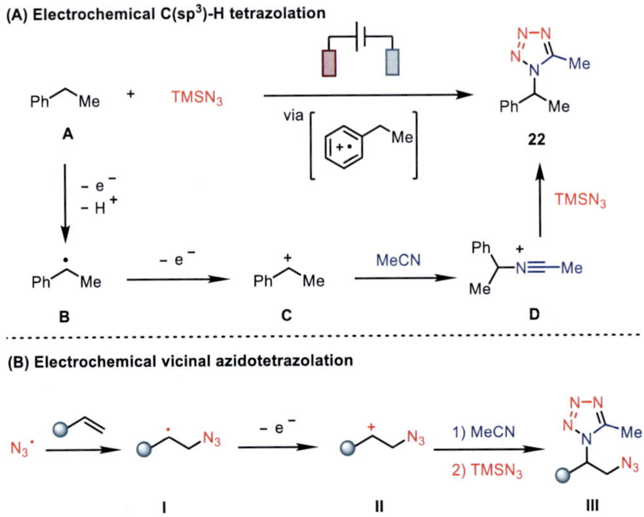

Scheme 19 Proposed mechanism for electrochemical $C(sp^3)$–H tetrazolation and azidotetrazolation

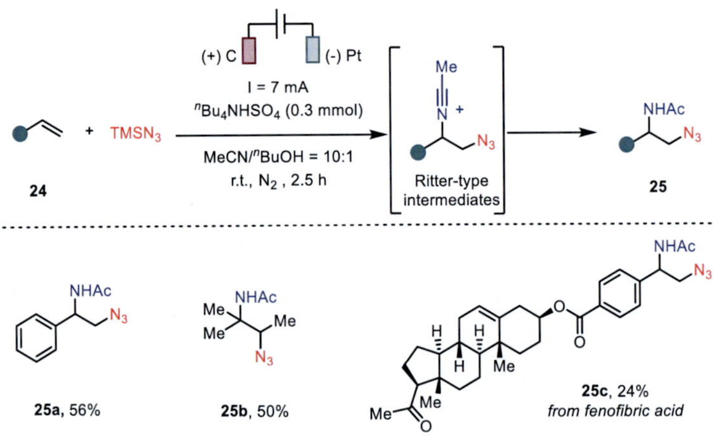

Scheme 20 Electrochemical synthesis of vicinal azidoacetamides

competing diazidation is disfavored for two reasons. On the one hand, the formation of vicinal diazide from the cross-coupling of two transient radicals, i.e., carbon-centered radical (**I**) and azido radical ($N_3^{\bullet}$), is unfavorable. On the other hand, the formation of vicinal diazide from the nucleophilic addition of azide (N_3^-) to the carbenium ion (**II**) fails to out-compete MeCN because of the solvent quantity of the latter nucleophile.

In addition, we developed an environmentally friendly and effective electrochemical method to obtain various vicinal azido acetamides (Scheme 20) [43]. It was

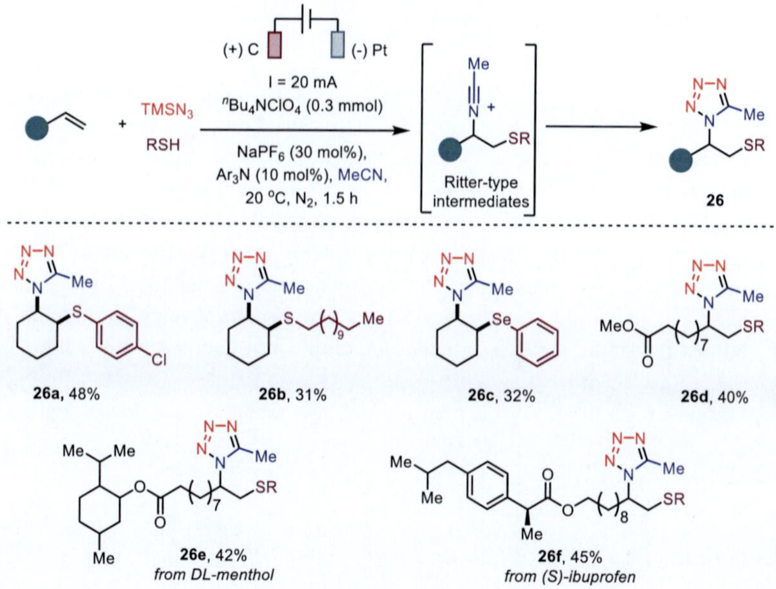

Scheme 21 Electrochemical vicinal sulfenyltetrazolation of unactivated alkenes

found that the electrolyte is an important parameter to tune product selectivity as the use of $^{n}Bu_4NBF_4$ afforded both the azidoacetamides and azidotetrazoles almost unselective, whereas $^{n}Bu_4NHSO_4$ favored the formation of azidotetrazoles but not azidotetrazoles. This method is simple in operation, features mild conditions, and provides a new synthetic approach to various diamine derivatives. The reaction tolerates both aryl (**25a**) and aliphatic alkenes (**25b**) and is compatible with the post-modification of complex molecules derived from fenofibric acid (**25c**).

We also developed an e-MCR with the readily available thiols, azidotrimethylsilane, and acetonitrile as a green and sustainable approach toward the sulfenyltetrazolation of unactivated alkenes, which are inaccessible with the existing methods (Scheme 21) [44]. It was also found that inorganic NaN_3 was much less effective in this transformation than that of azidotrimethylsilane. Various thiophenol (**26a**) and aliphatic thiols (**26b, 26d**) were well tolerated. This e-MCR was applied to the facile preparation of the analogous yet synthetically challenging vicinal selenotetrazole (**26c**) with diphenylselenide. Substrates derived from natural products and pharmaceutically active molecules, such as *DL*-menthol (**26e**) and (*S*)-ibuprofen (**26f**), were also viable.

4 Summary

The examples described in this chapter suggest that the electrochemical PCET strategy is a robust and versatile approach for the facile generation of nitrogen-centered radicals, which can be widely used in the synthesis of diverse nitrogen-containing heterocycles both via cascade cyclization and migration cyclization. In addition, the electrochemically generated Ritter-type intermediates also provide many new opportunities for the synthesis of various tetrazoles that are not accessible by other means. We anticipate these green and sustainable electrochemical approaches for the rapid construction of diverse *N*-heterocycles should find more applications in organic synthesis, medicinal chemistry, and materials sciences.

Acknowledgment Financial support from the National Natural Science Foundation of China (No. 22171046) is gratefully acknowledged.

References

1. Coetzee WA (2013) Multiplicity of effectors of the cardioprotective agent, diazoxide. Pharmacol Ther 140:167–175
2. Myznikov LV, Vorona SV, Zevatskii YE (2021) Biologically active compounds and drugs in the tetrazole series. Chem Heterocycl Compd 57:224–233
3. Nicolas C, Verny M, Giraud I, Ollier M, Rapp M, Maurizis J-C, Madelmont J-C (1999) New quaternary ammonium Oxicam derivatives targeted toward cartilage: synthesis, pharmacokinetic studies, and antiinflammatory potency. J Med Chem 42:5235–5240
4. Topolovčan N, Gredičak M (2021) Synthesis and stereoselective catalytic transformations of 3-hydroxyisoindolinones. Org Biomol Chem 19:4637–4651
5. Kerru N, Gummidi L, Maddila S, Gangu KK, Jonnalagadda SB (2020) A review on recent advances in nitrogen-containing molecules and their biological applications. Molecules 25: 1909–1949
6. Pennington LD, Moustakas DT (2017) The necessary nitrogen atom: a versatile high-impact design element for multiparameter optimization. J Med Chem 60:3552–3579
7. Beckwith ALJ, Centenary Lecture (1993) The pursuit of selectivity in radical reactions. Chem Soc Rev 22:143–151
8. Cheng X, Lei A, Mei T-S, Xu H-C, Xu K, Zeng C (2022) Recent applications of homogeneous catalysis in electrochemical organic synthesis. CCS Chem 4:1120–1152
9. Jiang Y, Xu K, Zeng C (2018) Use of electrochemistry in the synthesis of heterocyclic structures. Chem Rev 118:4485–4540
10. Darcy JW, Koronkiewicz B, Parada GA, Mayer JM (2018) A continuum of proton-coupled electron transfer reactivity. Acc Chem Res 51:2391–2399
11. Weinberg DR, Gagliardi CJ, Hull JF, Murphy CF, Kent CA, Westlake BC, Paul A, Ess DH, McCafferty DG, Meyer TJ (2012) Proton-coupled electron transfer. Chem Rev 112:4016–4093
12. Yu X-Y, Zhao Q-Q, Chen J, Xiao W-J, Chen J-R (2020) When light meets nitrogen-centered radicals: from reagents to catalysts. Acc Chem Res 53:1066–1083
13. Gentry EC, Knowles RR (2016) Synthetic applications of proton-coupled electron transfer. Acc Chem Res 49:1546–1556
14. Tarantino KT, Liu P, Knowles RR (2013) Catalytic Ketyl-olefin cyclizations enabled by proton-coupled electron transfer. J Am Chem Soc 135:10022–10025

15. Murray PRD, Cox JH, Chiappini ND, Roos CB, McLoughlin EA, Hejna BG, Nguyen ST, Ripberger HH, Ganley JM, Tsui E, Shin NY, Koronkiewicz B, Qiu G, Knowles RR (2021) Photochemical and electrochemical applications of proton-coupled electron transfer in organic synthesis. Chem Rev 122:2017–2291
16. Nguyen LQ, Knowles RR (2016) Catalytic C–N bond-forming reactions enabled by proton-coupled electron transfer activation of amide N–H bonds. ACS Catal 6:2894–2903
17. Shi Z, Li Y, Li N, Wang WZ, Lu HK, Yan H, Yuan Y, Zhu J, Ye KY (2022) Electrochemical migratory cyclization of *N*-acylsulfonamides. Angew Chem Int Ed 61:e202206058
18. Kornahrens AF, Cognetta AB, Brody DM, Matthews ML, Cravatt BF, Boger DL (2017) Design of benzoxathiazin-3-one 1,1-dioxides as a new class of irreversible serine hydrolase inhibitors: discovery of a uniquely selective PNPLA4 inhibitor. J Am Chem Soc 139:7052–7061
19. Chen N, Xu H-C (2021) Electrochemical generation of nitrogen-centered radicals for organic synthesis. Green Synth Catal 2:165–178
20. Fuentes N, Kong W, Fernández-Sánchez L, Merino E, Nevado C (2015) Cyclization cascades viaN-amidyl radicals toward highly functionalized heterocyclic scaffolds. J Am Chem Soc 137: 964–973
21. Kärkäs MD (2018) Electrochemical strategies for C–H functionalization and C–N bond formation. Chem Soc Rev 47:5786–5865
22. Zhao Q-Q, Hu X-Q, Yang M-N, Chen J-R, Xiao W-J (2016) A visible-light photocatalytic N-radical cascade of hydrazones for the synthesis of dihydropyrazole-fused benzosultams. Chem Commun 52:12749–12752
23. Shi Z, Dong S, Liu T, Wang W-Z, Li N, Yuan Y, Zhu J, Ye K-Y (2024) Electrochemical cascade migratory versus *ortho*-cyclization of 2-alkynylbenzenesulfonamides. Chem Sci 15:2827–2832
24. Zheng PC, Cheng J, Su S, Jin Z, Wang YH, Yang S, Jin LH, Song BA, Chi YR (2015) Oxidative N-heterocyclic carbene-catalyzed γ-carbon addition of Enals to imines: mechanistic studies and access to antimicrobial compounds. Chem A Eur J 21:9984–9987
25. Shi Z, Li N, Wang W-Z, Lu H-K, Yuan Y, Li Z, Ye K-Y (2022) Electrochemical 5-*exo-dig* aza-cyclization of 2-alkynylbenzamides toward 3-hydroxyisoindolinone derivatives. Org Biomol Chem 20:4320–4323
26. Shi Z, Wang W-Z, Li N, Yuan Y, Ye K-Y (2022) Electrochemical dearomative spirocyclization of N-acyl thiophene-2-sulfonamides. Org Lett 24:6321–6325
27. Shi Z, Lu H-K, Li N, Yuan Y, Li Z, Ye K-Y (2022) Electrochemical oxidative dearomatization of 2-arylthiophenes. Org Chem Front 9:2921–2925
28. Atobe M, Tateno H, Matsumura Y (2017) Applications of flow microreactors in electrosynthetic processes. Chem Rev 118:4541–4572
29. Elsherbini M, Wirth T (2019) Electroorganic synthesis under flow conditions. Acc Chem Res 52:3287–3296
30. Liu T, Shi Z, Yuan Y, Lin Y, Ye KY (2024) Electrochemical dearomative spirocyclization of N-acyl sulfonamides in a continuous-flow cell. Chin J Chem 42:980–984
31. Li N, Shi Z, Yuan Y, Li Z, Ye K-Y (2022) Rapid synthesis of spirodienones via electrochemical dearomative spirocyclization in flow. Org Chem Front 9:6586–6591
32. Chen C, Luo Y, Yin H, Zhong Q, Zheng S, Liu R, Zhao C, Wang G, He L (2022) Design, synthesis, and antitumor activity evaluation of novel acyl sulfonamide spirodienones. Biorg Med Chem 60:116626–116636
33. Jiang Y-M, Lin Y-Y, Zhu L, Yu Y, Li Y, Lin Y, Ye K-Y (2024) A general electrochemical synthesis of sulfonimidoyl fluorides, azides, and acetates. CCS Chem:1. https://doi.org/10.31635/ccschem.023.202303489
34. Ritter JJ, Kalish J (1948) A new reaction of nitriles. II. Synthesis of t-carbinamines. J Am Chem Soc 70:4048–4050
35. Dhiman N, Kaur K, Jaitak V (2020) Tetrazoles as anticancer agents: a review on synthetic strategies, mechanism of action and SAR studies. Biorg Med Chem 28:115599–115619

36. Myznikov LV, Hrabalek A, Koldobskii GI (2007) Drugs in the tetrazole series. (review). Chem Heterocycl Compd 43:1–9
37. Dippold AA, Izsák D, Klapötke TM, Pflüger C (2016) Combining the advantages of tetrazoles and 1,2,3-triazoles: 4,5-bis(tetrazol-5-yl)-1,2,3-triazole, 4,5-bis(1-hydroxytetrazol-5-yl)-1,2,3-triazole, and their energetic derivatives. Chem A Eur J 22:1768–1778
38. Wang T, Gao H, Shreeve JNM (2021) Functionalized tetrazole energetics: a route to enhanced performance. Z Anorg Allg Chem 647:157–191
39. Yuan Y, Lei A (2019) Electrochemical oxidative cross-coupling with hydrogen evolution reactions. Acc Chem Res 52:3309–3324
40. Meyer TH, Choi I, Tian C, Ackermann L (2020) Powering the future: how can electrochemistry make a difference in organic synthesis? Chem 6:2484–2496
41. Yan M, Kawamata Y, Baran PS (2017) Synthetic organic electrochemical methods since 2000: on the verge of a renaissance. Chem Rev 117:13230–13319
42. Yu Y, Zhu X-B, Yuan Y, Ye K-Y (2022) An electrochemical multicomponent reaction toward C–H tetrazolation of alkyl arenes and vicinal azidotetrazolation of alkenes. Chem Sci 13: 13851–13856
43. Yu Y, Yuan Y, Ye K-Y (2023) Electrochemical synthesis of vicinal azidoacetamides. Chem Commun 59:422–425
44. Zhu X-B, Yu Y, Yuan Y, Ye K-Y (2023) Electrochemical multicomponent reaction toward vicinal sulfenyltetrazolation of unactivated alkenes. Org Chem Front 10:5064–5069

Top Heterocycl Chem (2024) 61: 135–162
https://doi.org/10.1007/7081_2024_82

Published online: 27 September 2024

Electrochemically Mediated Construction and Functionalization of *N*/*S*-Heterocycles

Jiangwei Wen, Jianjing Yang, and Yonghong Yin

Contents

Abstract The utilization of electrochemistry for the construction and functionalization of *N*/*S*-heterocycles has found extensive application in a wide range of synthetic transformations. The focus of this chapter lies in the recent advancements in the methodology of electrochemically assisted radical generation and its subsequent application in the synthesis and functionalization of diverse *N*/*S*-heterocyclic compounds. The methodology and mechanism of electrochemically mediated synthesis and functionalization of *N*/*S*-heterocycles will be discussed in terms of direct anodic oxidation, indirect oxidation, cathodic reduction, and paired electrolysis.

Keywords Cathode reduction · Construction of *N*/*S*-heterocycles · Direct oxidation · Free radicals · Functionalization of *N*/*S*-heterocycles · Indirect oxidation · Paired electrolysis · Undivided cells

---

J. Wen (✉), J. Yang, and Y. Yin
Key Laboratory of Green Natural Products and Pharmaceutical Intermediates in Colleges and Universities of Shandong Province, School of Chemistry and Chemical Engineering, Qufu Normal University, Qufu, People's Republic of China
e-mail: wenjy@qfnu.edu.cn; jjyang@whu.edu.cn

1 Introduction

N/*S*-heterocyclic compounds are widely distributed in natural products, functional materials, and synthetic molecules, playing crucial roles in biological and industrial applications alike. Their profound significance is manifested in various aspects, with their biological activities being particularly noteworthy. The majority of pharmaceuticals and biologically active agrochemicals consist of heterocyclic compounds, as exemplified by the 2015 Nobel molecules *Artemisinin* and *Abamectin*, for example [1]. Consequently, organic chemists have been devoting significant efforts to the synthesis of these *N*/*S*-heterocyclic compounds through the development of novel and efficient synthetic methodologies. Transition-metal catalysis and free-radical chemistry have long played pivotal roles in the synthesis of *N*/*S*-heterocyclic compounds, representing indispensable approaches in this field [2–6]. In recent years, with a deepening understanding of green chemistry by synthetic chemists, a series of novel and interesting synthesis strategies such as photocatalysis [7], electrocatalysis [8–10], and heterogeneous catalysis [11] are emerging. In particular, the utilization of traceless electrons as sustainable redox agents for the synthesis of *N*/*S*-heterocyclic compounds has garnered significant attention among organic synthetic chemists in the last decade [12, 13]. During this period, our laboratory has conducted extensive research on the synthesis and modification of *N*/*S*-heterocyclic compounds, resulting in the development of a series of efficient, mild, and environmentally friendly protocols for their construction through electrochemical pathways. Given this progress, the goal of this chapter is to provide a summary of our laboratory's work on the synthesis and modification of *N*/*S*-heterocyclic compound by electrochemistry in an undivided cell. We hope to provide readers with an intuitive, systematic, and clear methodological understanding of the construction and functionalization of *N*/*S*-heterocycles by electrochemistry through the review of this chapter. The scope and limitations of these electrochemical reactions are discussed as well as the corresponding mechanisms.

2 Functionalization of *N*/*S*-Heterocycles by Electrochemistry

The functionalization of *N*/*S*-heterocyclic compounds provides a direct approach to access a diverse array of novel frameworks, avoids heterocycle prefunctionalization, and offers unique opportunities to endow distinctive biological activity to these compounds. In general, organic electrosynthesis is conducted by utilizing sustainable electrons as activation reagents for oxidative coupling at the anode surface and reductive coupling at the cathode surface. In this section, we will discuss the functionalization of *N*/*S*-heterocyclic compounds from two perspectives: anodic oxidation and cathodic reduction via a radical cross-coupling pathway (Scheme 1). Among them, anodizing involves both direct and indirect oxidation processes to

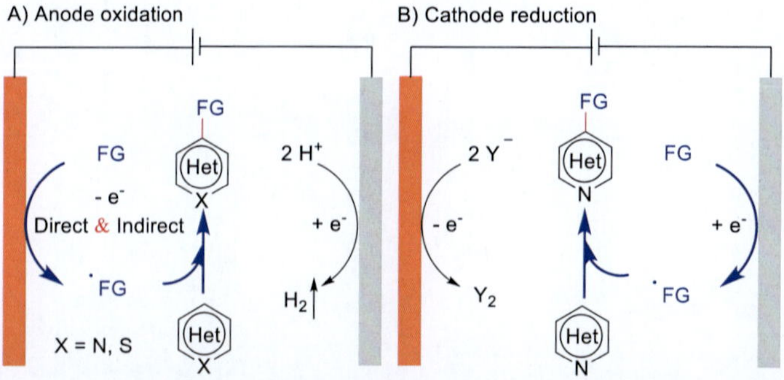

Scheme 1 Anode oxidation and cathode reduction strategies for the functionalization of *N*/*S*-heterocycles

generate the corresponding free radical intermediates, with concurrent hydrogen evolution reaction as the cathode half-reaction.

2.1 *Anodic Oxidation Mediates the Functionalization of* N/S*-Heterocycles*

The functionalization of ketene dithioacetals has been recognized as one of the most efficient approaches toward the construction of structurally diverse polysubstituted alkene molecules incorporating sulfur-containing heterocycles [14]. However, the conventional protocols for thiocyanation and alkylation of ketene dithioacetals still necessitate the utilization of transition-metal catalysts and stoichiometric chemical oxidants, posing challenges in reaction selectivity [15–17]. To address these problems, a novel and highly selective thiocyanation and alkylation of ketene dithioacetals regulated by water under constant current conditions were first disclosed [18]. It is noteworthy that this electrochemical protocol exhibits broad functional-group compatibility, including those bearing electron-withdrawing, electron-donating, alkyl, and heteroatomic substituents of ketene dithioacetals, leading to exceptional yields (Scheme 2). Under the optimized conditions, the scope of the electrochemical homo-coupling of ketene dithioacetals was investigated, and the results are summarized in Scheme 3. The obtained results are of great interest and surpass the achievements attained through previous methodologies. Furthermore, the reaction scope can be expanded to the carbon–nitrogen π bonds, wherein hydrazone **4** derived from an aldehyde smoothly underwent the thiocyanation process, resulting in product **5a** with a high yield of 95% under the optimized reaction conditions (Scheme 4). Based on the results of mechanistic experiments, we propose potential mechanisms for the thiocyanation and alkenylation of ketene dithioacetals, as depicted in Scheme 5. Initially, KSCN undergoes oxidation to form $(SCN)_2$ in the presence of H_2O, which is subsequently converted to thiocyanyl radical **Int-1**. Subsequently, the desired thiocyanation of ketene dithioacetals is

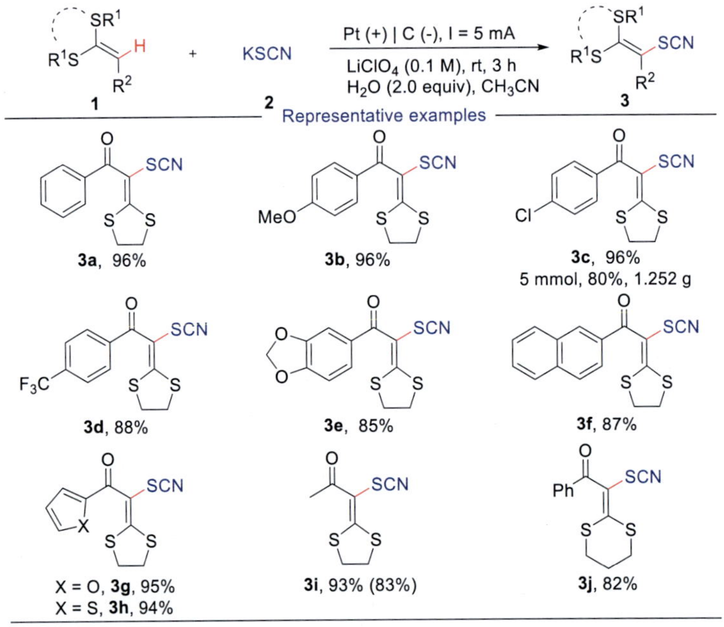

Scheme 2 Thiocyanation of ketene dithioacetals

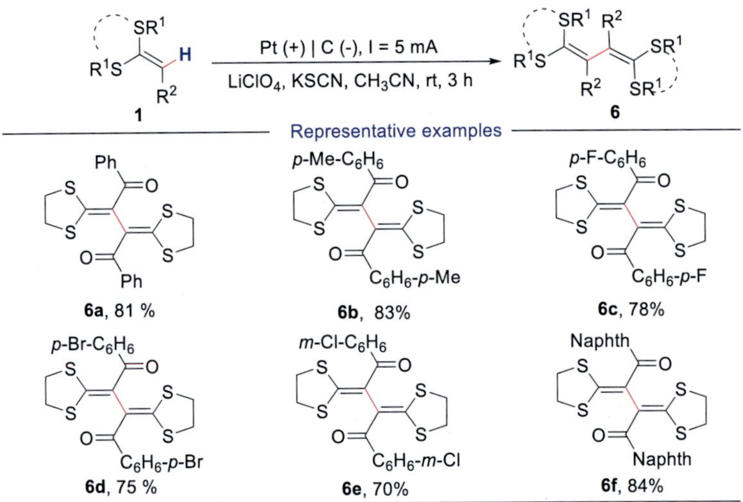

Scheme 3 Homodimerization of ketene dithioacetals

delivered via a radical addition and deprotonation process via the intermediate **Int-2**. In the absence of water, the delivery of thiocyanated products **3a** proceeds through a radical nucleophilic pathway and subsequently undergoes dethiocyanation in a halogen-like manner to yield compound **6a**.

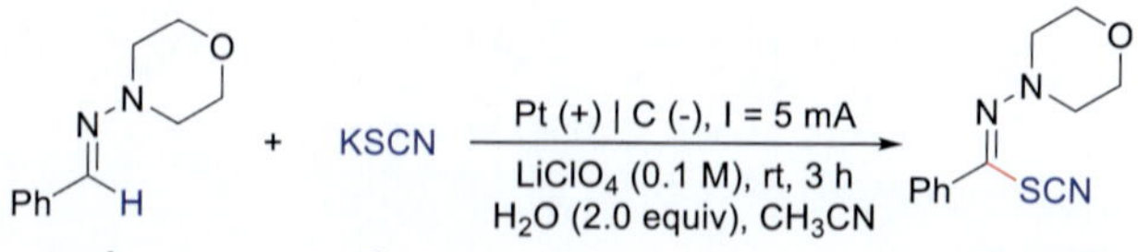

Scheme 4 The thiocyanation of hydrazones

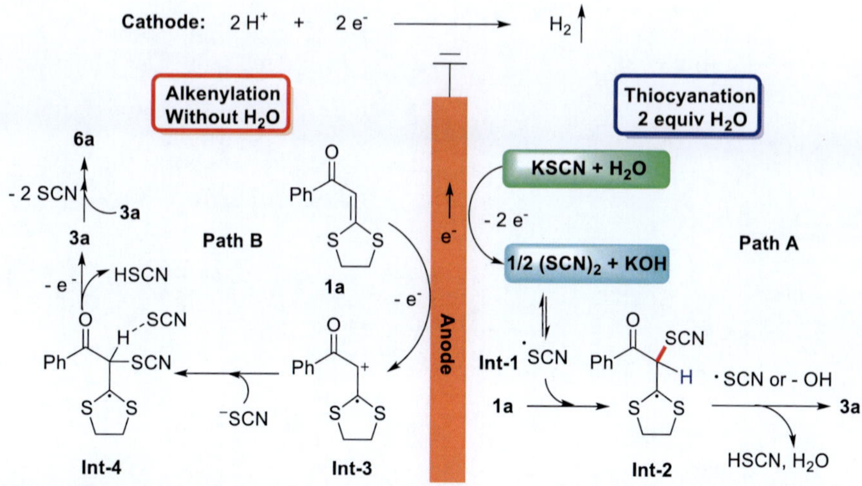

Scheme 5 A possible mechanism for the thiocyanation and alkenylation of ketene dithioacetals

Imidazo[1,2-a]pyridines, being recognized as pivotal scaffolds among *N*-heterocyclic compounds, have garnered significant attention from the scientific community due to their diverse biological activities encompassing antiviral, antitumor, antiparasitic, and antimicrobial properties alongside their fungicidal potential [19–22]. Consequently, due to the special structural features of imidazopyridine rings, extensive efforts have been dedicated to the direct C3–H functionalization of imidazo[1,2-a]pyridines using transition-metal or stoichiometric-oxidant-mediated approaches, such as alkoxylation, thiocyanation, sulfenylation, sulfonation, phosphorylation, carboxylation, and amination. Among them, the employment of thiocyanate derivatives as a stable sulfur source for achieving the thiomethylation of imidazopyridines has received limited attention under electrochemical conditions. In 2019, Lei [23] and coworkers successfully achieved C-3 sulfenylation of imidazopyridines with thiols under electrochemical conditions. However, the application of thiophenol in this method is constrained due to its high cost, pungent odor, and inherent instability. To address these limitations, a three-component cross-coupling strategy involving electrochemically mediated thiomethylation of imidazopyridines has been reported (Scheme 6) [24]. In this approach, thiocyanate was employed as an inexpensive, stable, and odorless sulfur source while methanol served as the methyl reagent. Notably, this scalable methodology operates without

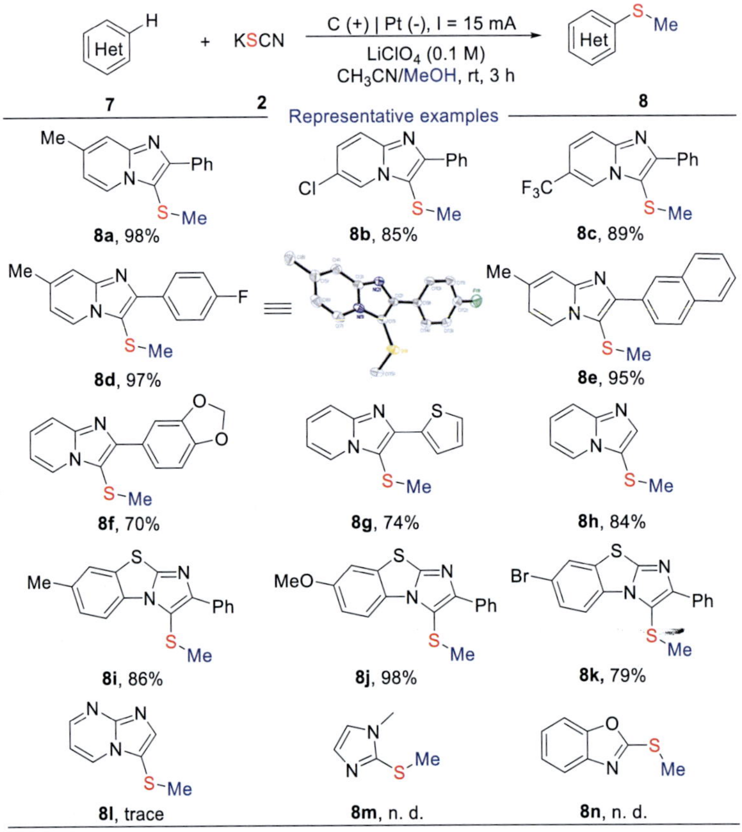

Scheme 6 Electrochemical-induced C3 thiomethylation of imidazo[1,2-a]pyridine

the need for metal catalysts or exogenous chemical oxidants with excellent gram-scale yields. This electrochemical protocol enables efficient C3 thiomethylation of imidazo[1,2-a]pyridine and can be successfully extended to benzo[d]imidazo[2,1-b] thiazole. However, the current electrochemical conditions exhibit certain limitations in terms of substrate diversity, as compounds **8l-8n**, for example, cannot be obtained under established conditions.

The transformation's mechanism was comprehensively investigated through a series of meticulously designed control and deuteration experiments. Firstly, deuteration experiments demonstrated that the methyl group of the product originates from methanol (Scheme 7a). Experiments subjecting intermediates to electrolysis have substantiated that thiocyanation plays a pivotal role in the current transformation (Scheme 7b). The absence of any desired product in the free-radical inhibition experiment thereby indicated that the transformation should proceed via a radical mechanism (Scheme 7c). Finally, based on the experimental findings, a plausible mechanism was proposed for the regioselective C3 thiomethylation of imidazopyridines with KSCN and methanol under the electrochemical conditions,

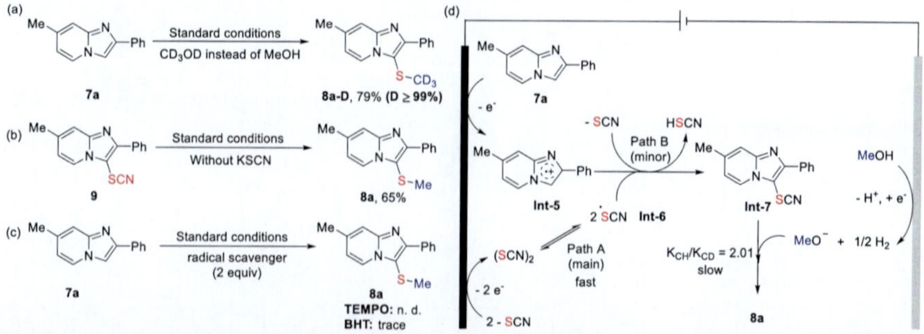

Scheme 7 Key mechanism studies and possible mechanisms

as depicted in Scheme 7d. The substrate **7a** undergoes initial single-electron oxidation at the anode, resulting in the formation of intermediate **Int-5**. In the meantime, KSCN undergoes oxidation on the anode surface, leading to the formation of $(SCN)_2$, which is in equilibrium with thiocyanyl radical **Int-6** that undergoes efficient cross-coupling with intermediate **Int-5**. Finally, intermediate **Int-7** reacts with methanol to furnish the desired product **8a** under the basic conditions, while H_2 is released as the sole byproduct at the cathode. Kinetic-isotopic-effect (KIE) experiments ($K_{CH}/K_{CD} = 2.01$) suggested that the catalytic cycle may involve the dissociation of methanol as the rate-determining step.

Di(indolyl)methanes (DIMs), as highly significant scaffolds, are widely found in various pharmaceutical and natural products, such as Vibrindole A, Arsindoline A, Turbomycin B, Trisindoline, and Arundine, exhibiting intriguing anticancer, antifungal, antimicrobial, antibacterial, and antitumor biological activities [25–32]. In light of this, numerous sophisticated methodologies have been established for the synthesis of DIM frameworks through the powerful transition-metal-catalyzed Friedel-Crafts alkylation of indoles with alcohols as alkylating reagents [27, 33–41]. Despite significant advancements, the construction of DIM molecules remains challenging using existing C–H alkylation methods, likely due to the higher energy barrier ($\Delta G = 40.9$ kcal mol^{-1}) associated with Friedel-Crafts alkylation of indoles with alcohols. The direct electrolysis-enabled C–H alkylation of indoles with alcohols for the synthesis of di(indolyl)methanes was initially documented in 1994 [42]. After a lapse of 20 years, Ghanbari [43] developed a direct approach for synthesizing DIMs in undivided cells through the implementation of segmented potentiostatic electrolysis. These direct oxidation processes necessitated the utilization of a divided cell or high potential (2.0 V, then −0.9 V vs. SCE), and the scope was limited. Indirect electrolysis, by employing redox catalysts as electron shuttles, offers advantages in terms of mitigating electrode passivation and kinetic inhibition, while also enabling enhanced selectivity and energy efficiency. Therefore, a metal-free approach was disclosed for the unified synthesis of diverse DIMs through *N*-hydroxyphthalimide-*N*-oxyl radical (NIPO)-catalyzed electrochemical C–H alkylation of indole derivatives (Scheme 8) [44]. In terms of the substrate scope, various indoles and primary alcohols were compatible with the present protocol,

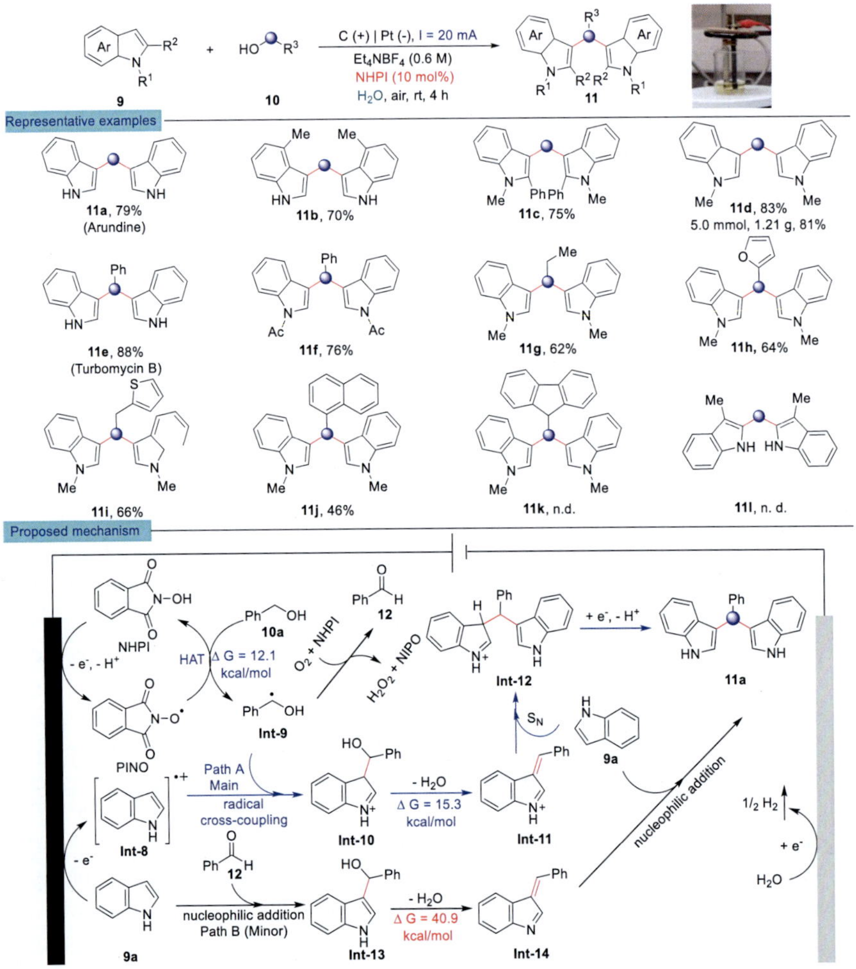

Scheme 8 Indirect electrolysis access to di(indolyl)methanes

generating 40 desired products in up to 88% yield with water as a green solvent. Based on the experiments and DFT calculations, a possible reaction mechanism was proposed, as shown in Scheme 8. Initially, the radical intermediate NIPO generated by the single-electron oxidation of NHPI undergoes hydrogen-atom transfer (HAT) with benzyl alcohol (**10a**) to give carbon radical intermediate **Int-9** ($\Delta G = 12.1$ kcal mol^{-1}). Concurrently, indole (**9a**) is also anodized, resulting in the formation of aryl radical cation intermediate **Int-8**. The cationic intermediate **Int-10** was generated through the free radical cross-coupling of the carbon radical intermediate **Int-9** with the aryl radical cationic intermediate **Int-8**.

Subsequently, intermediate **Int-11** is readily obtained via dehydration of intermediate **Int-10** ($\Delta G = 15.3$ kcal mol^{-1}). Finally, the desired product **11a** was obtained by H-atom abstraction from intermediate **Int-12**, which was generated by the nucleophilic addition of another indole to intermediate **Int-11** (path A, main). In contrast, the DFT calculations reveal that the conventional Friedel-Crafts mechanism pathway assumes a subordinate role (path B, $\Delta G = 40.9$ kcal mol^{-1}).

Pyrazolones, as one of the highly valuable and widespread scaffolds, are present in numerous natural products, medicinal, and agrochemical industries due to their diverse pharmacological and biological activities. The pyrrole core is included in some commercially available drugs such as Celebrex, Lexiscan, Fipronil, Crizotinib, and Edaravone [45–52]. Among these, edaravone functions as an antioxidant and scavenger of free radicals, primarily employed in the clinical management of cerebral hemorrhage and cerebral infarction. In particular, tripyrazolones featuring edaravone as the core have garnered significant attention from the pharmaceutical industry due to their potential for qualitative analysis of edaravone. However, conventional synthetic approaches for these heterocyclic compounds are constrained by elevated temperatures, stoichiometric metal oxidants, and challenges associated with residual metals [53–55]. In 2021, an electrochemical/I^-induced dual-catalysis strategy was reported for the three-component $C(sp^3)$-H dehydrogenative trimerization of pyrazolones, enabling access to tripyrazolones (Scheme 9) [56]. The electrochemical oxidative strategy described herein can be readily scaled up for the synthesis of biologically active tripyrazolones, thereby facilitating their practical applications. A plausible mechanism for the electrochemical/I^-induced dual-catalysis three-component dehydrogenative trimerization of pyrazolones has been proposed based on the experimental data and previous literature. Moreover, sulfonated pyrazoles can be delivered by employing an electrochemical/I^- dual-catalysis strategy under external oxidant- and metal-free conditions. The established electrochemical/I^- dual-catalyzed strategy exhibits remarkable substrate versatility, efficiency, and scalability to afford gram quantities of desired products (Scheme 10a) [57]. Based on experimental data from mechanistic studies and relevant literature reports, three potential reaction pathways were inferred, as illustrated in Scheme 10b. I^- undergoes a two-electron oxidation process at the anode to form I^+, which reacts with **12** to generate the crucial intermediate **Int-22**, and then undergoes nucleophilic substitution with sodium sulfinate **14**, which may represent the fastest pathway for product formation (Path A). However, the significance and feasibility of paths B and C should not be disregarded. The anodically generated iodonium ions can undergo catalytic cycling, being reoxidized on the surface of the anode, while H^+ will be reduced on the cathode surface to generate hydrogen gas as an environmentally friendly byproduct.

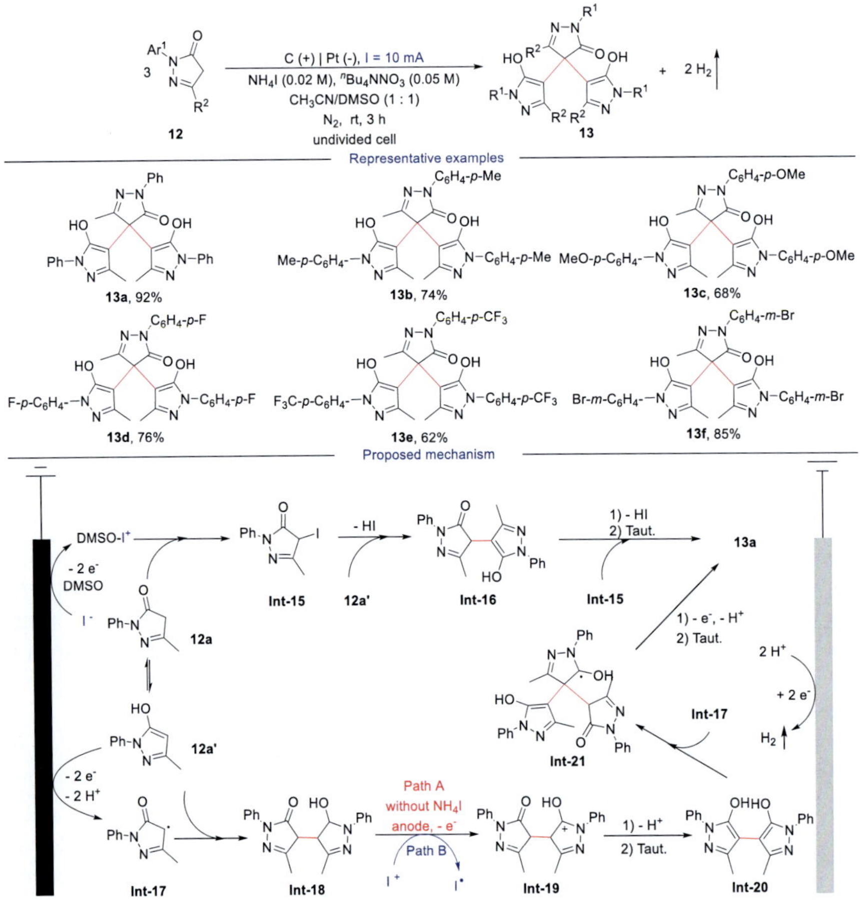

Scheme 9 Dehydrogenative trimerization of pyrazolones to tripyrazolones

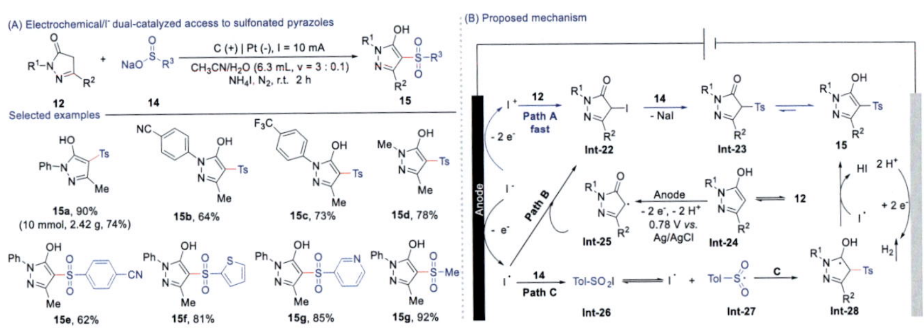

Scheme 10 Electrochemical/I⁻ dual-catalyzed access to sulfonated pyrazoles

2.2 *Cathodic Reduction Mediates the Functionalization of* N-*Heterocycles*

In the previous section, we discussed how anodic oxidation mediates the functionalization of *N*/*S*-heterocycles, and various direct and indirect oxidation strategies are included. In this section, we will preliminarily summarize typical cases wherein cathodic reduction mediates the functionalization of *N*-heterocycles. Although anodic oxidation provides a powerful tool for the selective functionalization of *N*-heterocyclic compounds, the cathodic reduction of H^+ and subsequent hydrogen evolution during anodizing processes is frequently overlooked in the literature [58]. As we all know, $H_2 = 2e^- + 2H^+$ has been recognized as a sustainable and environmentally friendly alternative to traditional hydrogen. Consequently, the utilization of electrons as activation reagents and H^+ as a hydrogen source presents a unique opportunity for selective hydrogenation reduction of *N*-heterocyclic compounds on the cathode surface, thereby circumventing the necessity for high-pressure hydrogen and transition-metal catalysts [12].

In the preceding section, we discussed the significance of imidazopyridine *N*-heterocyclic compounds and their C3 thiomethylation. In recent years, imidazopyridines have emerged as prominent *N*-heterocyclic compounds due to their distinctive structure, facilitating a range of C3 functionalization reactions by electrochemical anodic oxidation. Nevertheless, the cathodic transfer hydrogenation of imidazopyridine remains a challenging task in current research. The model reaction substrate chosen for this study was 7-methyl-2-phenylimidazo[1,2-a]pyridine (**7a**), and its reduction potential in various solvents was evaluated by cyclic voltammetry (CV) experiments [59]. To our delight, the adoption of a mixed solvent of CH_3CN/THF (9:1) exhibited significant reductive current at a synthetically accessible potential, indicating the pronounced reducing ability of this electrochemical system (Scheme 11a). The validity of these CV conclusions can be confirmed by electrosynthetic experiments, wherein the desired product **17a** could be obtained in up to 88% yields (Scheme 11b, entry 1).

(A) Investigation of solvent effect of **1a** in cathode

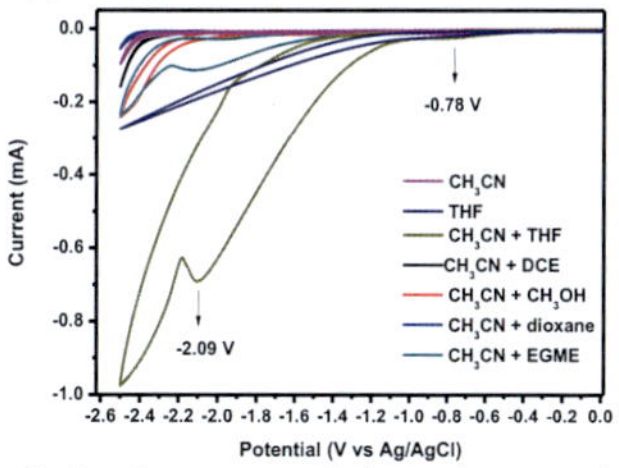

Cyclic voltammograms at glass carbon as work electrode, Pt (1 x 1 cm) as counter electrode in 0.1 M $^nBu_4NBF_4$ with different solvent under the N_2, **7a** (0.25 mM) , the mixed solvent of CH_3CN/others (v = 9 :1), scan rate 100 mV/s.

(B) Experimental verification

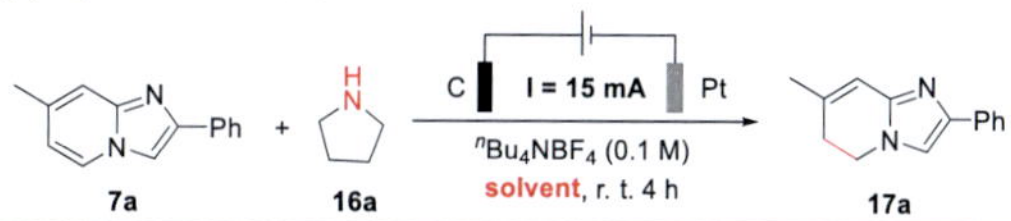

Entry	Solvent	Yield (%)[b]
1	CH_3CN/THF	88
2	CH_3CN	35
3	THF	23
4	CH_3CN/EGME	46
5	CH_3CN/dioxane	trace
6	CH_3CN/DCE	trace
7	CH_3CN/CH_3OH	trace

Scheme 11 CV and electrolysis experiments of **7a**

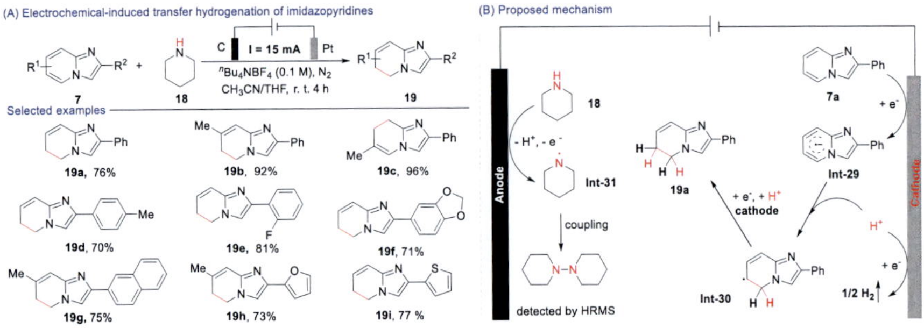

Scheme 12 Electrochemical-induced transfer hydrogenation of imidazopyridines

With the optimized conditions in hand, the substrate scope was investigated to assess the utility and versatility of the electrochemically induced transfer hydrogenation of imidazopyridines with piperidine. Under these electrochemical conditions, a series of imidazopyridine derivatives exhibited remarkable efficacy in transfer hydrogenation, displaying exceptional yield and selectivity (Scheme 12a). These findings underscore their significant potential for efficient hydrogenation in organic synthesis. Based on the preliminary experimental findings and existing literature reports, a reasonable mechanism has been proposed for the selective transfer hydrogenation of imidazopyridines by using piperidine as the hydrogen donor (Scheme 12b). The substrate **7a** initially undergoes single-electron reduction on the surface of the Pt cathode, leading to the formation of an intermediate radical anion **Int-29**, which is protonated to afford the radical intermediate **Int-30**. The latter intermediate is subsequently subjected to further single-electron reduction at the cathode surface, yielding the desired product **19a** after protonation. At the anode, piperidine, serving as a hydrogen donor, undergoes oxidative coupling to generate byproduct 1,1′-bipiperidine.

The construction of functional N-containing active biomolecules and bidentate nitrogen ligands by electroreductive pyridinylation of *N*-heteroaromatics represents a captivating task and challenge in the field. In particular, quinolin-2(1H)-ones, as a class of privileged *N*-heterocyclic compounds, are widely used in numerous natural products, medicines, pesticides, and functional materials [60–66]. Therefore, an electrochemically mediated proton-coupled electron transfer strategy was developed to synthesize bidentate nitrogen ligands and *N*-heterocycles from cheap and readily available cyanopyridines and quinolinones as starting materials (Scheme 13a) [67]. According to experiments and previous reports, a probable reaction mechanism was proposed for the electroreductive C3 pyridinylation of quinoxalin-2(1H)-ones with cyanopyridines (Scheme 13b). Initially, the generation of free radical intermediate **Int-33** was facilitated through a proton-coupled electron transfer (PCET) process from **21a** in the presence of hexafluoroisopropanol (HFIP) at the cathode surface. Subsequently, radical intermediate **Int-33** and **20a** undergo radical addition to obtain *N*-centered radical intermediate **Int-34**, which was subjected to anodization

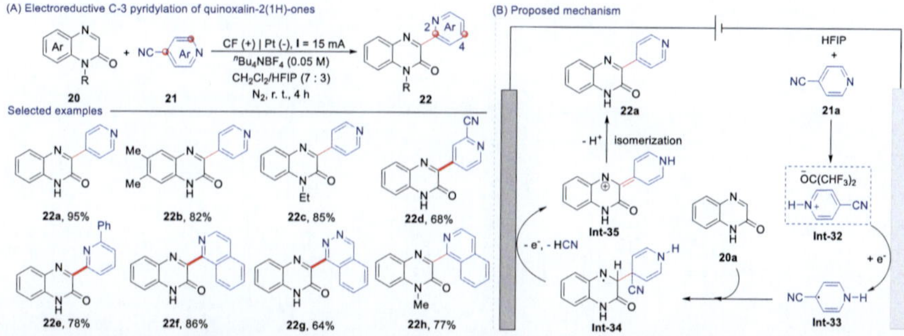

Scheme 13 Electroreductive C-3 pyridylation of quinoxalin-2(1H)-ones

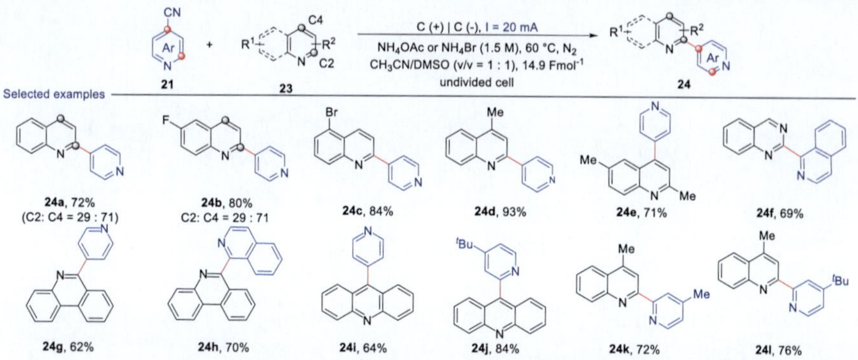

Scheme 14 Electrochemical ammonium-cation-assisted pyridylation of inert *N*-heterocycles

and decyanation to generate *N*-centered cation **Int-35**. Finally, desired product **22a** was obtained by deprotonation of intermediate **Int-35**.

Pyridines are widely recognized as prominent heterocycles in pharmaceuticals, materials, natural product molecules, and organic functional materials. It is widely acknowledged that electron-deficient pyridine derivatives are frequently converted into cationic salts, which serve as efficient radical acceptors for Minisci reactions, whereas the neutral heterocycles are insufficiently electrophilic to undergo radical addition. Analogously, owing to their inherently negative electrode potentials in their neutral forms, they are challenging to activate by cathodic reduction [68]. The electrochemically mediated radical cross-coupling reaction between cyanopyridine with inert *N*-heterocycles thus presents a direct but challenging approach to accessing biologically active *N*-heterocyclic compounds. To address this issue, a novel electrochemical NH_4^+-assisted pyridinylation of the inert *N*-heterocycles approach has been developed through a dual-PCET and radical cross-coupling mechanism facilitated by sequentially paired electrolysis [69]. A variety of important *N*-fused heterocycles and bidentate nitrogen ligand compounds can be obtained with moderate to excellent yields under these electrochemical conditions (Scheme 14). Interestingly, the resulting product **24i** can serve as a highly sensitive fluorescence

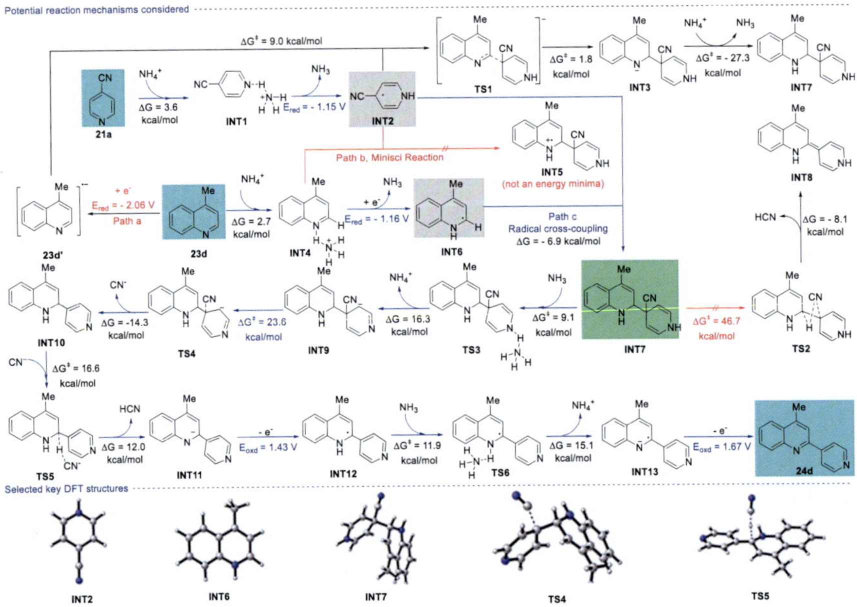

Scheme 15 Consideration of mechanistic scenarios and DFT evaluated

sensor for the recognition of Fe^{2+} and Pd^{2+}, with an impressive detection limit of 2.5×10^{-5} mmol/L for Fe^{2+}.

To gain a deeper understanding of the dual PCET-mediated free-radical coupling mechanism, a comprehensive DFT calculation was conducted (Scheme 15). Based on these calculations, the reduction potentials of **23d** and **21a** on the surface of the cathode have been confirmed to be remarkably challenging, reaching potentials of −2.06 V and −1.6 V, respectively. However, upon complexation with NH_4^+, their reduction potentials are moderated to −1.16 V and −1.15 V, respectively. These findings suggest that the electrochemically mediated dual PCET strategy represents the optimal pathway for generating crucial free radical intermediates **INT-2** and **INT-6**. The classical Minisci reaction mechanism has been excluded by DFT calculations.

This electrochemical dual PCET strategy was then employed to accomplish the hydropyridinylation of diverse α,β-unsaturated carbonyl compounds with good yields in an undivided cell, encompassing chalcone, arimistane, megestrol acetate, ketene dithioacetal, chromone, and other electrophilic alkenes (Scheme 16) [44]. The present electrochemical NH_4^+-assisted dual PCET strategy features a broad substrate scope and allows gram-scale synthesis. The versatile role of NH_4Cl as a hydrogen donor, protonation reagent, and electrolyte should be underscored in this transformation. According to a series of CV results, it was observed that the reduction potential difference between the α,β-unsaturated ketone and cyanopyridine in an NH_4^+-assisted dual PCET system is within ~0.4 V for reaction initiation (Scheme 16a). Moreover, the mechanism of dual PCET and

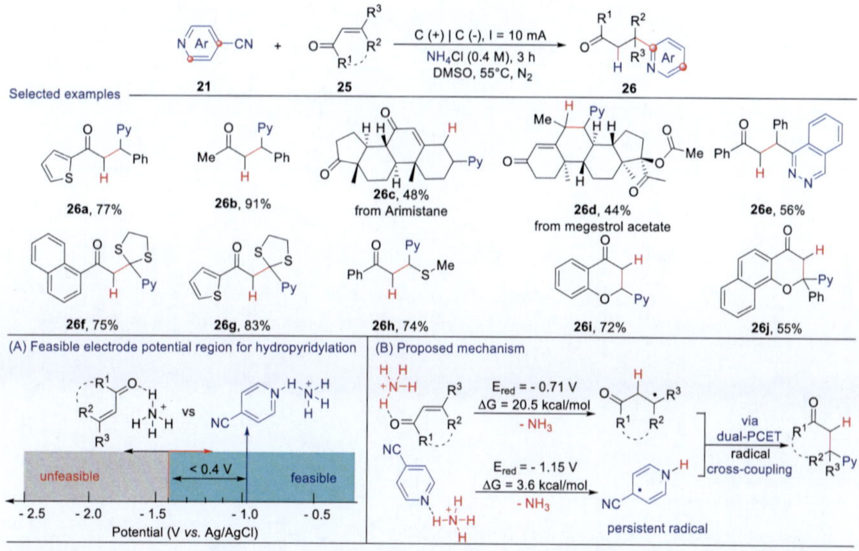

Scheme 16 Electrochemical NH_4^+-assisted hydropyridinylation of ketone-activated alkenes

radical cross-coupling has been elucidated by a comprehensive analysis of experimental evidence and DFT calculations (Scheme 16b).

Precise control of electron transfer for site-selective hydropyridinylation of ynones for the synthesis of β-pyridyl ketones and pyridinated propargyl alcohols is appealing, yet it remains a formidable challenge. Square-wave voltammetry (SWV) techniques were employed to accurately and rapidly identify the electron transfer processes involved in the hydropyridinylation of ynones, thereby facilitating the elucidation of the reaction mechanism. To our delight, SWV experiments confirmed that ynones undergo a three-electron transfer in dimethyl sulfoxide (DMSO) and a single-electron transfer in methylene chloride (DCM, CH_2Cl_2). These divergent, solvent-mediated factors govern site selectivity (Scheme 17) [44]. The present method exhibits controllable electron transfers, exceptional regioselectivity, facile gram-scale synthesis, and environmentally benign characteristics.

Based on experimental observations, a plausible mechanistic pathway for site-selective hydropyridinylation of ynones was proposed, as depicted in Scheme 18. We posit that the judicious selection of solvent and protonation reagents plays a pivotal role in governing the site selectivity in the hydropyridylation of ynones. With DMSO as the solvent, **27a** is reduced through a net reduction involving $3e^-$ and $3H^+$ to yield radical intermediate **Int-38**, which subsequently undergoes radical cross-coupling and decyanation with free radical intermediate **Int-41** to afford desired product **28a**. The reduction of **21a** and **27a** occurs simultaneously at the cathode via a dual PCET process in dichloromethane (DCM), resulting in the generation of free radical intermediates **Int-42** and **Int-41**. Subsequently, these intermediates undergo radical cross-coupling and decyanation processes, ultimately leading to the formation of product **29a**.

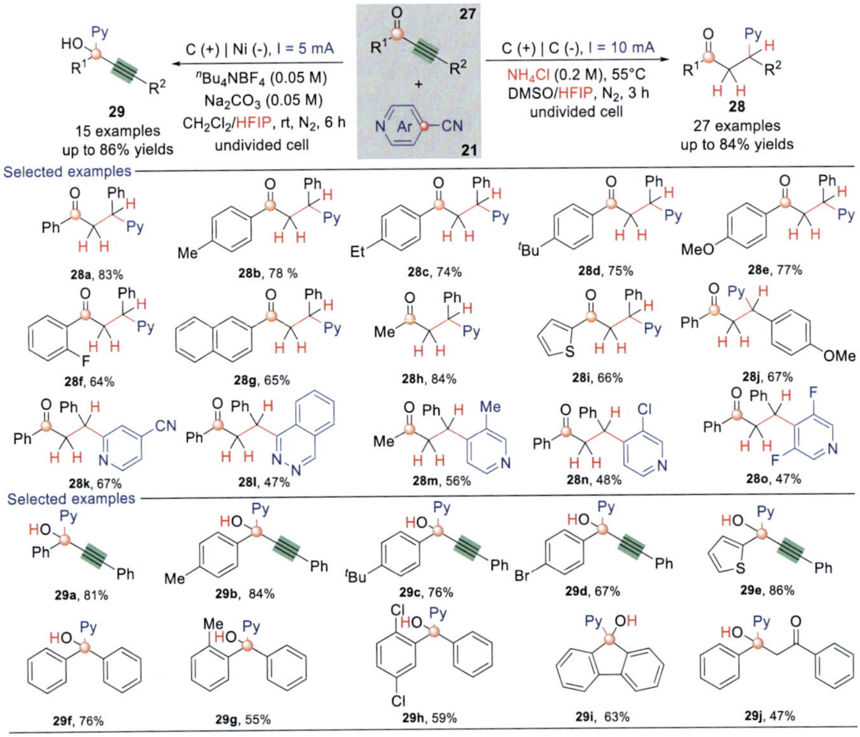

Scheme 17 Site-selective hydropyridylation of ynones

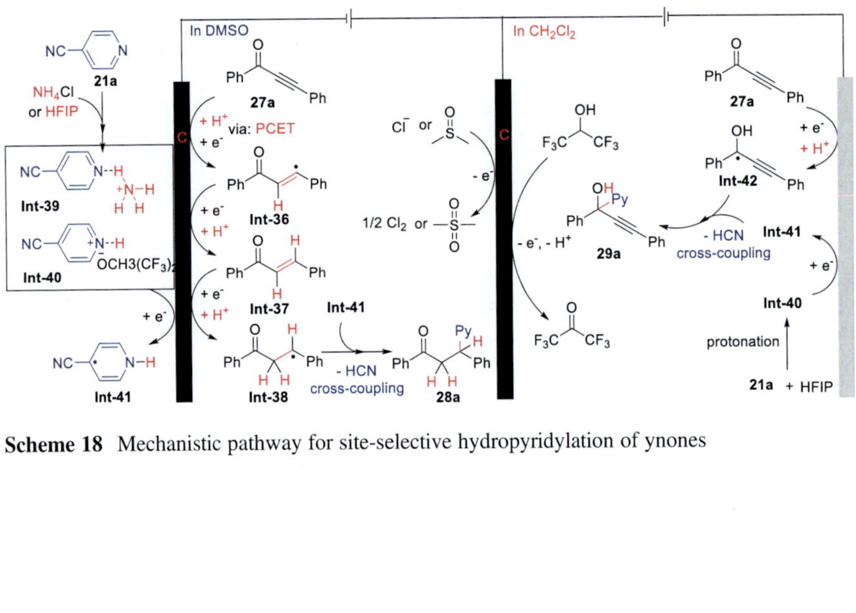

Scheme 18 Mechanistic pathway for site-selective hydropyridylation of ynones

3 Construction of *N/S*-Heterocycles by Electrochemistry

Thioimidazoles and thiazoles are privileged *N/S*-heterocyclic scaffolds frequently encountered in a natural products, pharmaceuticals, agrochemicals, and functional materials. As shown in Scheme 19, these representative bioactive compounds encompass thioimidazole and thiazole scaffolds, which exhibit high biological and pharmacological activity [70–74]. Therefore, the development of efficient, cost-effective, and environmentally sustainable methods for constructing these bioactive molecular scaffolds from inexpensive and readily available raw materials has garnered significant attention.

3.1 Construction of N/S*-Heterocycles by Paired Electrolysis*

In 2022, we presented a facile and efficient electrochemical approach for the synthesis of alkylthioimidazoles through cross-coupling reactions of isothiocyanates and isocyanates with nonactivated alkyl halides in the presence of water as a cosolvent. In terms of substrate scope, a wide range of primary, secondary, and tertiary unactivated haloalkanes and isothiocyanates can be effectively employed in this protocol, achieving 36 examples with yields as high as 92% (Scheme 20) [75]. The structures of **33i** (CCDC: 2098994) and **33u** (CCDC: 2098995) were unambiguously confirmed by X-ray single crystal diffraction. The one-pot, three-component cross-coupling strategy for paired electrolysis-induced thioalkylation can be readily scaled up, thus holding great promise for broader applications in the practical industrial field.

Mechanistic experiments have confirmed that this one-pot, three-component cross-coupling reaction undergoes a radical process mediated by paired electrolysis

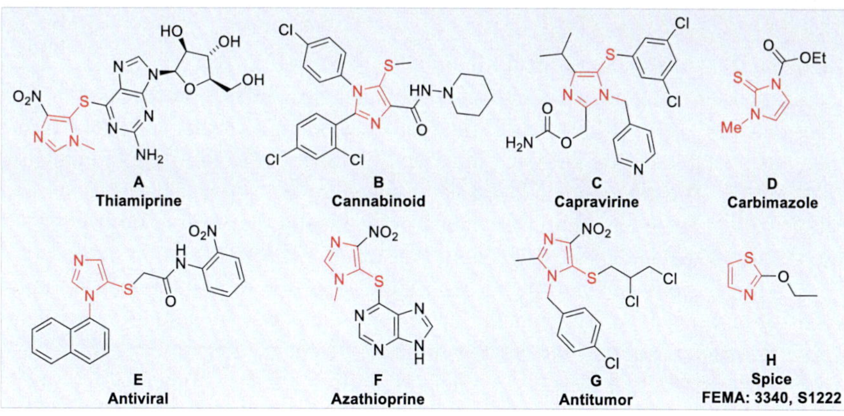

Scheme 19 Representative bioactive compounds encompass thioimidazoles and thiazoles

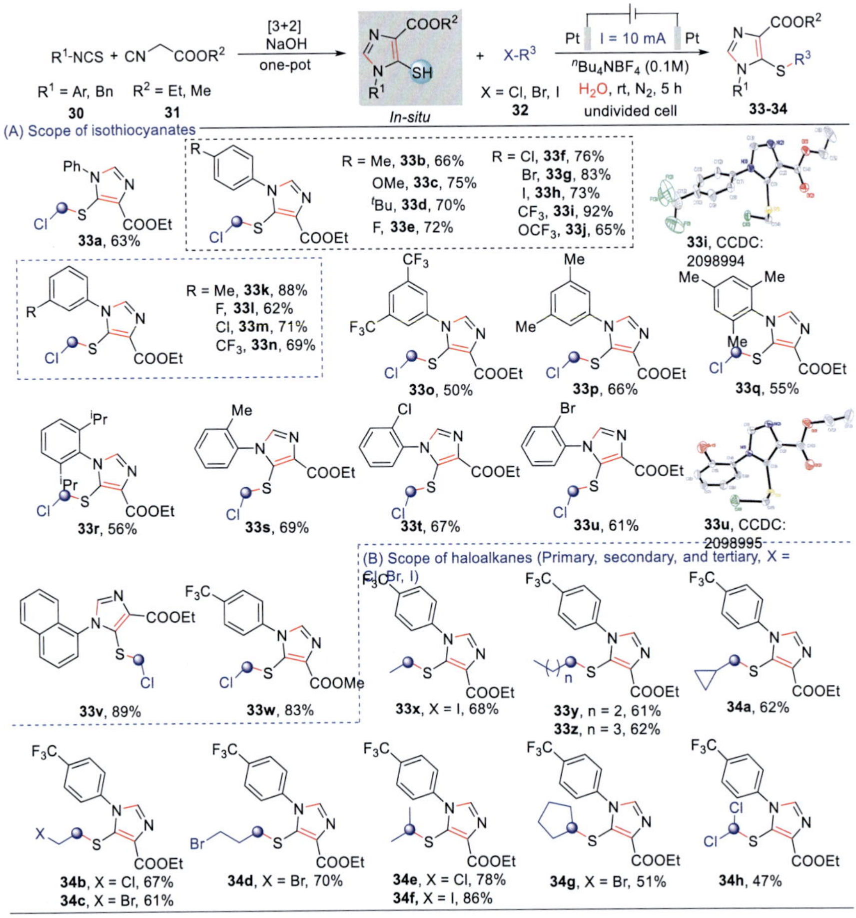

Scheme 20 Paired electrolysis mediated to access alkylthioimidazoles

(Scheme 21). Intermediate **Int-43** is initially generated in situ through the [3 + 2] cyclization of isothiocyanate **30** and isocyanate **31** in the presence of a base. Subsequently, intermediate **Int-43** undergoes oxidation on the anode surface to generate unstable radical **Int-44**, which rapidly dimerizes to form disulfide intermediate **Int-45**. This intermediate is further reduced on the cathode surface, generating sulfur radical **Int-44** and thiolate anion **Int-46**. Finally, the desired products **33–34** are delivered via a radical cross-coupling or nucleophilic-substitution process.

Electrosynthesis enables precise control over reaction energy through the adjustment of current or voltage, thereby achieving selective regulation. Chloroalkanes serve as crucial building blocks in synthesis; however, their direct activation by redox chemistry is constrained by their prohibitive reduction potentials. To tackle this challenge, we successfully achieved controllable radical coupling of thiophenol with dichloromethane by employing our consecutive paired electrolysis strategy to

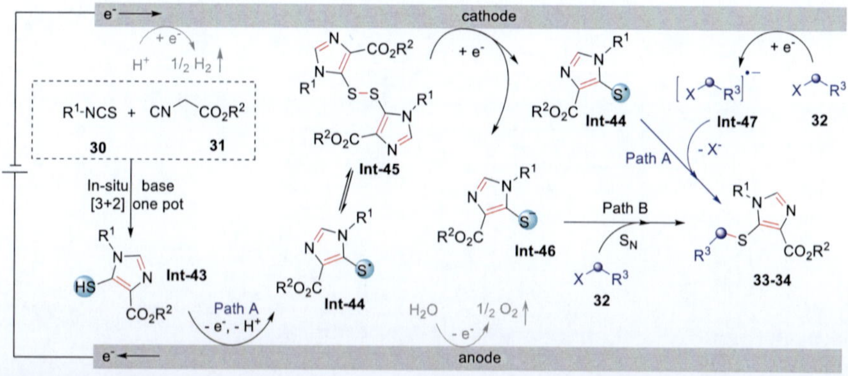

Scheme 21 Proposed mechanism for the synthesis of alkylthioimidazoles

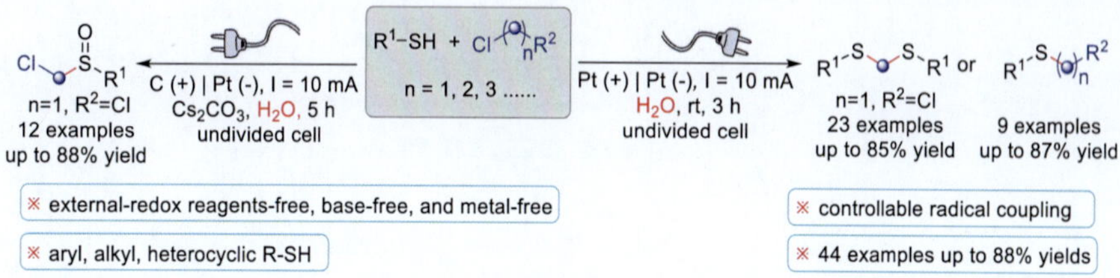

Scheme 22 Controllable cross-coupling of thiophenol with dichloromethane

deliver dithioacetals, sulfides, and sulfoxides in the absence of an external mediator (Scheme 22) [76]. The methodology offers wide substrate compatibility, straightforward operation, gram-scale synthesis, and environmental friendliness.

3.2 *Construction of* N/S*-Heterocycles by Anodic Oxidation*

Recently, tertiary amines with reduced nucleophilicity have gained significant attention as viable alternatives to highly reactive primary or secondary amines to donate amino groups through selective C–N cleavage [77–82]. Therefore, various 2-aminobenzoselenazoles can be synthesized by employing electrooxidation-induced, selective cleavage of C–N bonds of tertiary amines starting from 1-iodo-2-isoselenocyanatobenzenes and tertiary amines (Scheme 23) [83].

The development of novel, environmentally friendly, and pragmatic pesticide chemicals and bioactive molecules represents a prominent domain within the realm of organic synthetic chemistry. In recent years, over 860 pesticide chemicals have been documented globally, while the availability of commercial pesticides remains limited and exceedingly challenging [84–93].

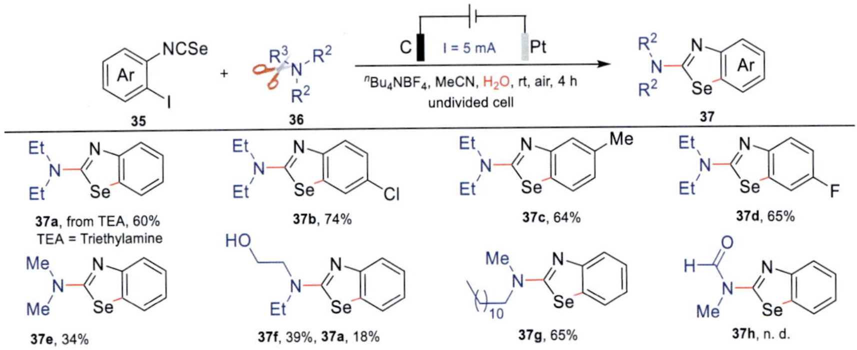

Scheme 23 Electrooxidation-induced selective cleavage of C–N bonds of tertiary amines to access 2-aminated benzoselenazoles

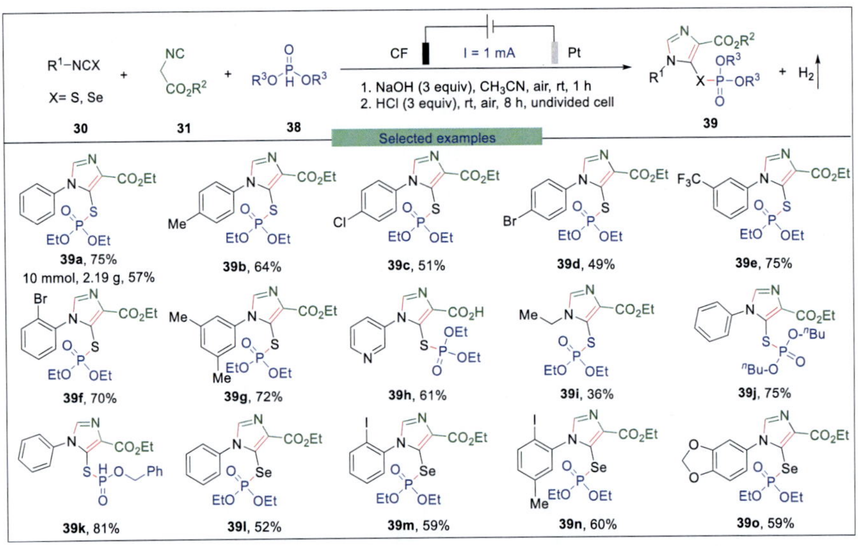

Scheme 24 Electrooxidation-induced synthesis 3-thio/selenophosphorylated imidazoles

Over an extended duration, persistent locust infestations have posed significant agricultural threats, necessitating the development of better pesticides. We anticipate that syntheses of novel thiophosphorylated imidazole derivatives through the combination of active profiles from thiophosphates and thiimidazoles could potentially deliver biocompatible pesticides. Therefore, we presented the first convenient and practical electrochemical method for accessing 3-thio/selenophosphorylated imidazoles from readily available isothiocyanates/isoselenocyanates, isocyanates, and H-phosphonates via a one-pot, three-component radical cross-coupling pathway (Scheme 24) [94]. In terms of the substrate scope, a wide range of isothiocyanates/isoselenocyanates and H-phosphonates can be employed in this protocol, forming

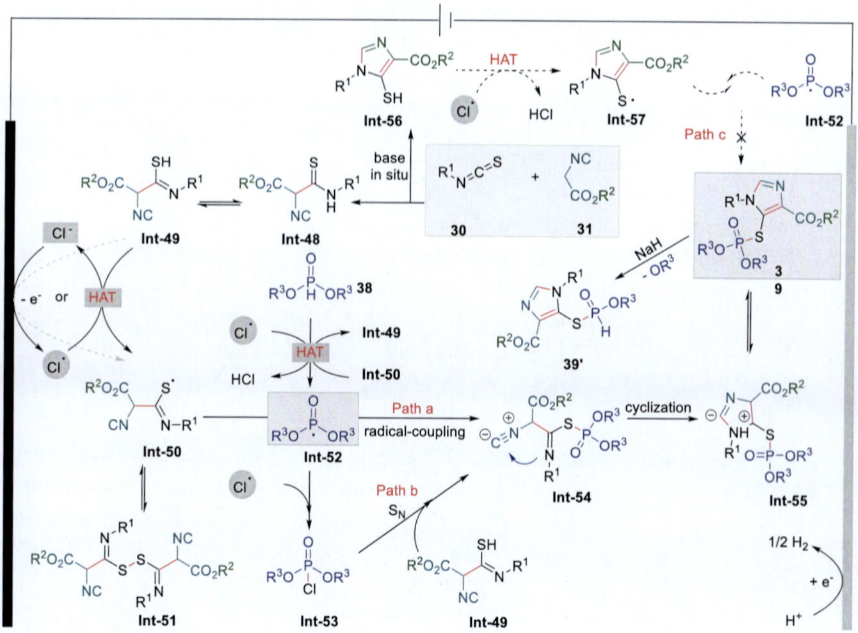

Scheme 25 Postulated reaction pathway

desired products with yields up to 81% and concurrent hydrogen release. Importantly, the readily accessible compounds can be utilized for antilocust treatments in wheat with good biocompatibility (RAW 264.7 macrophages, IL-6 = 5.3 pg/mL, THF-α = 79.8 pg/mL), highlighting the significance of this electrochemical methodology.

Notably, mechanistic studies have ruled out classical [3 + 2] cyclization and radical cross-coupling processes (path c), an alternative electrochemically mediated radical coupling, instead implicating cyclization pathways (path a or b). As depicted in Scheme 25, the intermediate **Int-49** are initially produced through in situ nucleophilic addition of isocyanates (**31**) to isothiocyanates (**30**) in the presence of the base. Next, the key free radical intermediates **Int-50** and **Int-52** are generated through hydrogen atom transfer (HAT) between intermediates **Int-49** and **38** with chlorine radicals, derived in turn from the anodic oxidation of chloride ions. The intermediate **Int-54** can be obtained through two pathways: one involved radical cross-coupling of radical intermediates **Int-50** and **Int-52** (path a, main), while the other pathway includes nucleophilic substitution of **Int-49** to **Int-53**, delivering **Int-54** in the presence of base (path b, minor). Finally, the desired product **39** is produced through spontaneous intramolecular cyclization of intermediate **Int-54**. Byproduct **39′** is obtained via selective removal of the OR^3 group, involving reaction of NaH generated from in situ cathodic reduction.

1,2,4-Triazolidines represent a distinctive class of saturated three-dimensional (3D) *N*-heterocycles that exhibit remarkable cytotoxicity against human

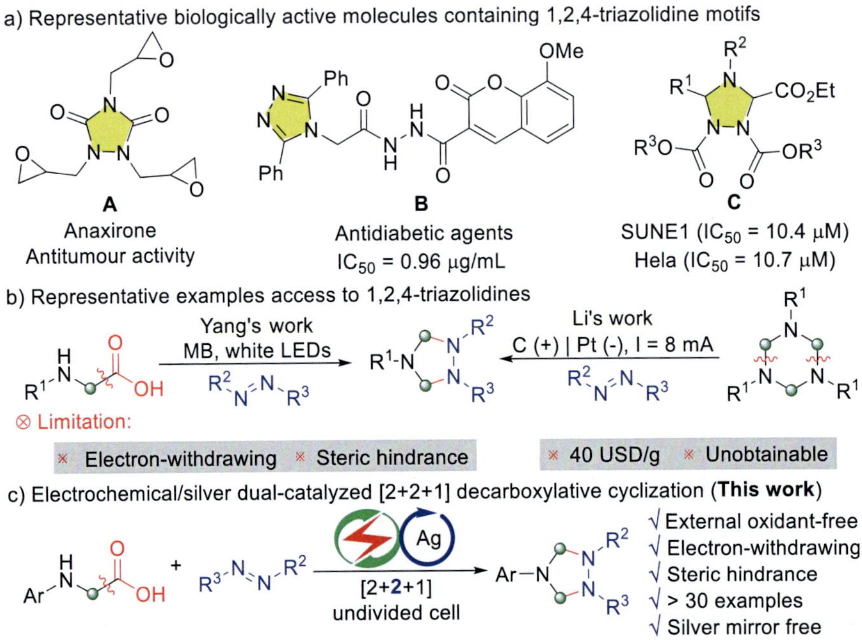

Scheme 26 Relevant biologically active molecules incorporating 1,2,4-triazolidines and their synthesis methods

nasopharyngeal carcinomas, tumors, diabetes, and cervical carcinomas (Scheme 26a) [95–97]. Hence, it is imperative to develop efficient and environmentally friendly synthetic methodologies for constructing this molecular framework utilizing cost-effective and readily accessible raw materials. However, the substrates utilized in previous elegant photocatalysis and electrosynthesis methods are limited by electron-withdrawing and steric-hindrance requirements, while C1 units remain prohibitively expensive (Scheme 26b) [98, 99].

To address these challenges, we developed a novel approach for the synthesis of 1,2,4-triazolidines through a silver-catalyzed decarboxylative [2 + 2 + 1] reaction of *N*-arylglycines with azobenzenes in undivided cells, utilizing constant current (Scheme 26c) [100]. Notably, this simple electrochemical silver-catalyzed decarboxylative cyclization method not only overcomes the limitations of previous methods, but it also ensures the catalytic activity of silver in undivided cells without forming silver mirrors. To our delight, this electrochemical protocol enables the decarboxylation of *N*-arylglycines bearing electron-withdrawing and sterically demanding groups and various azobenzenes through radical addition/cyclization, thereby facilitating access to 1,2,4-triazolidines (Scheme 27). The mechanism of this electrochemical silver-catalyzed [2 + 2 + 1] decarboxylation/coupling/cyclization has been substantiated through diverse experimental methodologies. These mechanistic studies demonstrated that an electrochemical/Ag^{II}-catalyzed radical addition/cyclization pathway is the predominant reaction mechanism (path a), while a base-

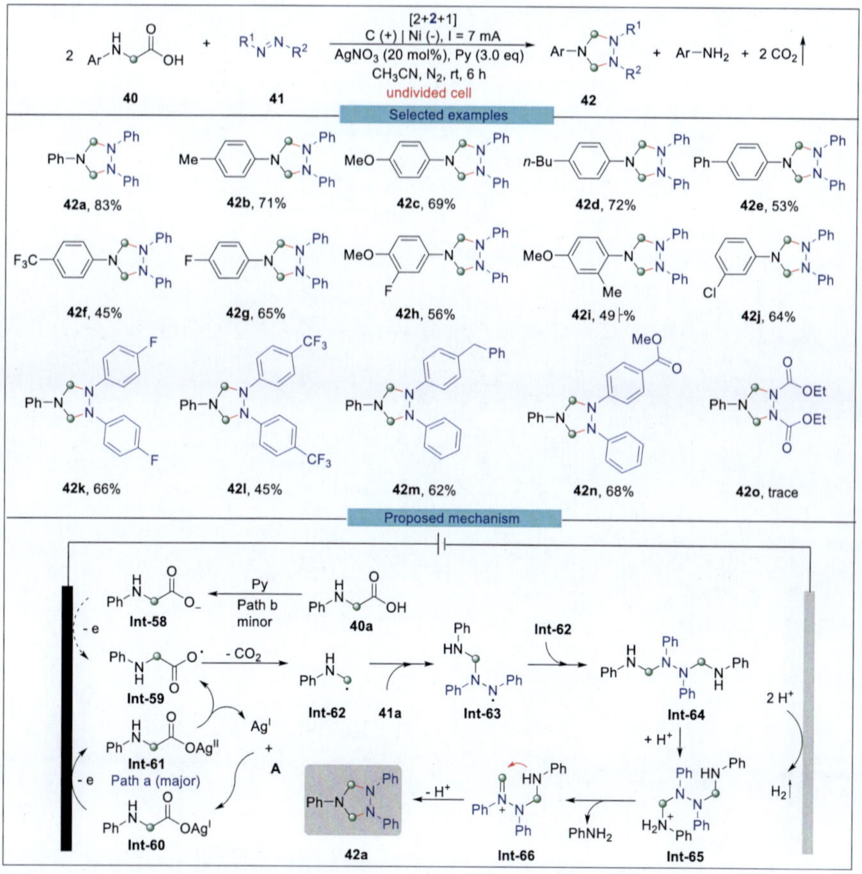

Scheme 27 Electrochemical silver-catalyzed decarboxylation access to 1,2,4-triazolidines

facilitated direct anodic oxidation pathway assumes a secondary role (path b). Additionally, these studies also unveiled that the use of *N*-arylglycines and pyridines in this electrochemical/Ag^{II}-mediated decarboxylation protocol plays a pivotal role in inhibiting the formation of silver mirrors.

Functionalized thiazoles have garnered significant attention owing to their intriguing biological activities, and thc site-selective incorporation of oxyalkyl groups onto these frameworks can profoundly modulate their pharmacological profiles [101, 102]. The diverse range of biological properties and applications associated with these compounds has motivated the development of novel strategies for achieving 2-oxoalkylation of thiazole frameworks. To overcome the limitations of traditional synthetic methods, such as unavailable starting materials, prefunctionalization requirements, and limited substrate scopes, we have developed an efficient electrochemical strategy for synthesizing 2-alkoxythiazole derivatives via a three-component cascade reaction in undivided cells.

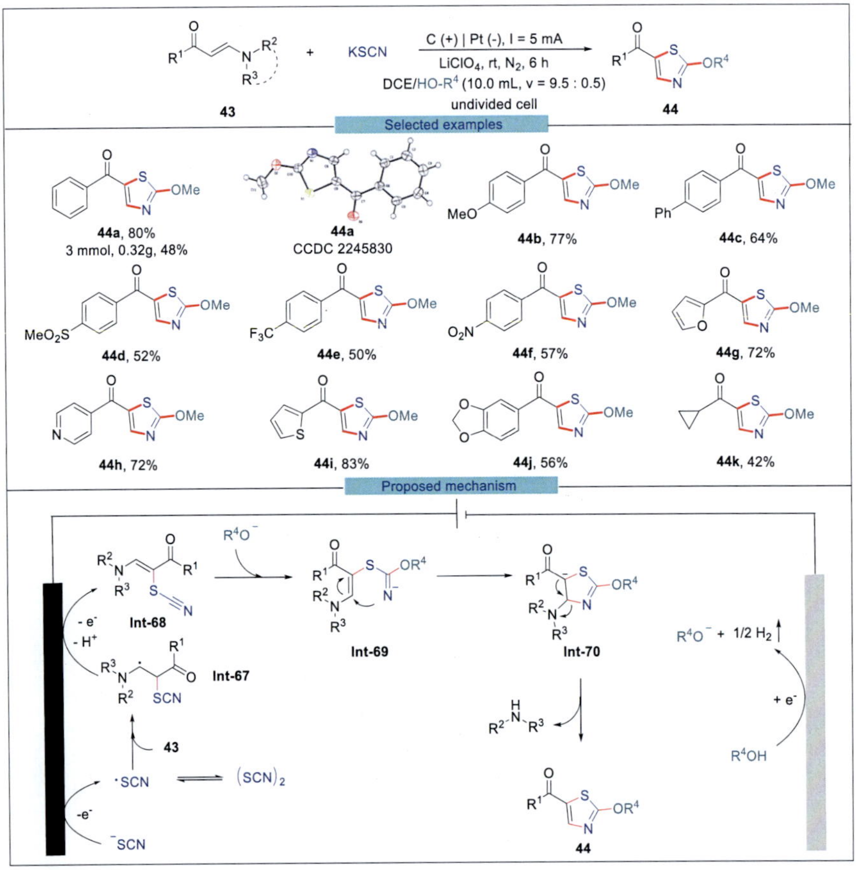

Scheme 28 Electrochemical synthesis of 2-alkoxyethiazole and proposed reaction pathway

The scalable electrochemical strategy described herein can accommodate a diverse range of electron-donating, electron-withdrawing, heteroatomic, and alkyl-substituted enaminones, enabling the synthesis of their corresponding products with yields ranging from moderate to excellent (Scheme 28) [103]. A plausible mechanism for the three-component cascade reaction is depicted in Scheme 28 on the basis of experimental results. The reaction initiates with the oxidation of the thiocyanate anion at the anode, leading to the formation of the thiocyanyl radical. Subsequently, this radical undergoes addition to enaminone **43**, generating carbon-centered radical intermediate **Int-67**, which is subsequently subjected to oxidation and dehydrogenation processes, yielding intermediate **Int-68**. Next, the thiocyanated enaminone undergoes attack by the alkoxy anion generated from dehydrogenation and reduction of alcohol at the cathode, leading to the formation of intermediate **Int-69**. Ultimately, desired product **44** is obtained through intramolecular nucleophilic cyclization and subsequent elimination of secondary amines.

4 Summary and Outlook

In this chapter, we have presented a comprehensive overview of our research group's endeavors in the electrochemical synthesis and functionalization of *N*/*S*-heterocyclic compounds. The construction and functionalization of *N*/*S*-heterocyclic compounds by electrochemistry are primarily discussed in terms of anode direct oxidation, indirect oxidation, cathode reduction, and paired electrolysis. Although these electrochemical strategies can achieve the synthesis of various small molecule *N*/*S*-heterocyclic compounds at the fundamental research level, there is still a significant gap in terms of their practical application in industrial production. Therefore, there is an urgent necessity for the development of efficient and environmentally sustainable electrochemical synthesis methods with industrial prospects, specifically targeting bioactive pesticides and pharmacologically active heterocyclic compounds. Excitingly, in recent years, novel organic electrochemical synthesis techniques have also been flourishing, offering a plethora of opportunities for organic electrochemical synthesis including flow electrochemistry, alternating current, and photoelectrochemistry. These methodologies have demonstrated significant potential for industrial application potentials and benefits in the fabrication of pharmacologically active pharmaceutical macromolecules. The incorporation of machine learning and artificial intelligence in the electrochemical synthesis of heterocyclic chemistry is also a prominent development trend, enabling efficient and precise construction as well as screening of drug molecules.

References

1. Tambo E, Khater EIM, Chen J-H, Bergquist R, Zhou X-N (2015) Infect Dis Poverty 4:58–64
2. Sharma UK, Sharma N, Vachhani DD, Van der Eycken EV (2015) Chem Soc Rev 44:1836–1860
3. Zhang M (2009) Adv Synth Catal 351:2243–2270
4. Guo X-X, Gu D-W, Wu Z, Zhang W (2015) Chem Rev 115:1622–1651
5. Yu J-T, Pan C (2016) Chem Commun 52:2220–2236
6. Landais Y (2018) https://doi.org/10.1007/978-3-319-89521-5
7. Chen J-R, Hu X-Q, Lu L-Q, Xiao W-J (2016) Acc Chem Res 49:1911–1923
8. Yamamoto K, Kuriyama M, Onomura O (2020) Acc Chem Res 53:105–120
9. Aslam S, Sbei N, Rani S, Saad M, Fatima A, Ahmed N (2023) ACS Omega 8:6175–6217
10. Swaroop TR, Narasimhamurthy KH, Umashankara M, Rangappa KS (2023) ChemistrySelect 8:e202302453
11. Bhaskaruni SVHS, Maddila S, Gangu KK, Jonnalagadda SB (2020) Arab J Chem 13:1142–1178
12. Yang J, Qin H, Yan K, Cheng X, Wen J (2021) Adv Synth Catal 363:5407–5416
13. Tay NES, Lehnherr D, Rovis T (2022) Chem Rev 122:2487–2649
14. Negishi E-I, Huang Z, Wang G, Mohan S, Wang C, Hattori H (2008) Acc Chem Res 41:1474–1485
15. Yu H, Jin W, Sun C, Chen J, Du W, He S, Yu Z (2010) Angew Chem Int Ed 49:5792–5797
16. Zhu N, Wang F, Chen P, Ye J, Liu G (2015) Org Lett 17:3580–3583

17. Chen Q, Lei Y, Wang Y, Wang C, Wang Y, Xu Z, Wang H, Wang R (2017) Org Chem Front 4:369–372
18. Wen J, Zhang L, Yang X, Niu C, Wang S, Wei W, Sun X, Yang J, Wang H (2019) Green Chem 21:3597–3601
19. Langtry HD, Benfield P (1990) Drugs 40:291–313
20. Lhassani M, Chavignon O, Chezal J-M, Teulade J-C, Chapat J-P, Snoeck R, Andrei G, Balzarini J, De Clercq E, Gueiffier A (1999) Eur J Med Chem 34:271–274
21. Enguehard-Gueiffier C, Gueiffier A (2007) Mini-Rev Med Chem 7:888–899
22. Gallud A, Vaillant O, Maillard LT, Arama DP, Dubois J, Maynadier M, Lisowski V, Garcia M, Martinez J, Masurier N (2014) Eur J Med Chem 75:382–390
23. Yuan Y, Cao Y, Qiao J, Lin Y, Jiang X, Weng Y, Tang S, Lei A (2019) Chin J Chem 3:49–52
24. Wen J, Niu C, Yan K, Cheng X, Gong R, Li M, Guo Y, Yang J, Wang H (2020) Green Chem 22:1129–1133
25. Gillespie Doreen E, Brady Sean F, Bettermann Alan D, Cianciotto Nicholas P, Liles Mark R, Rondon Michelle R, Clardy J, Goodman Robert M, Handelsman J (2002) Appl Environ Microbiol 68:4301–4306
26. Safe S, Papineni S, Chintharlapalli S (2008) Cancer Lett 269:326–338
27. Shiri M, Zolfigol MA, Kruger HG, Tanbakouchian Z (2010) Chem Rev 110:2250–2293
28. Bharate SB, Bharate JB, Khan SI, Tekwani BL, Jacob MR, Mudududdla R, Yadav RR, Singh B, Sharma PR, Maity S, Singh B, Khan IA, Vishwakarma RA (2013) Eur J Med Chem 63:435–443
29. Sun X-X, Du B-X, Zhang H-H, Ji L, Shi F (2015) ChemCatChem 7:1211–1221
30. Zhang M-Z, Chen Q, Yang G-F (2015) Eur J Med Chem 89:421–441
31. Dong Y, Lushnikova T, Golla RM, Wang X, Wang G (2017) Biorg Med Chem 25:864–869
32. Lee J (2019) Nutr Cancer 71:992–1006
33. Jella RR, Nagarajan R (2013) Tetrahedron 69:10249–10253
34. Gopalaiah K, Chandrudu S, Devi A (2015) Synthesis 47:1766–1774
35. Kaswan P, Nandwana NK, DeBoef B, Kumar A (2016) Adv Synth Catal 358:2108–2115
36. Pu F, Li Y, Song Y-H, Xiao J, Liu Z-W, Wang C, Liu Z-T, Chen J-G, Lu J (2016) Adv Synth Catal 358:539–542
37. Zheng L, Gao F, Yang C, Gao GL, Zhao Y, Gao Y, Xia W (2017) Org Lett 19:5086–5089
38. Qiao C, Liu X-F, Fu H-C, Yang H-P, Zhang Z-B, He L-N (2018) Chem Asian J 13:2664–2670
39. Yang T, Lu H, Shu Y, Ou Y, Hong L, Au C-T, Qiu R (2020) Org Lett 22:827–831
40. Jin J, Li Y, Xiang S, Fan W, Guo S, Huang D (2021) Org Biomol Chem 19:4076–4081
41. Dipika SYB, Pant S, Dhaked DK, Guru MM (2022) Org Chem Front 9:3428–3437
42. Suda K, Takanami T (1994) Chem Lett 23:1915–1916
43. Nikoofar K, Ghanbari K (2015) Monatsh Chem 146:2021–2027
44. Li B, Qin H, Yan K, Ma J, Yang J, Wen J (2022) Org Chem Front 9:6861–6868
45. Terrett NK, Bell AS, Brown D, Ellis P (1996) Bioorg Med Chem Lett 6:1819–1824
46. Penning TD, Talley JJ, Bertenshaw SR, Carter JS, Collins PW, Docter S, Graneto MJ, Lee LF, Malecha JW, Miyashiro JM, Rogers RS, Rogier DJ, Yu SS, Anderson GD, Burton EG, Cogburn JN, Gregory SA, Koboldt CM, Perkins WE, Seibert K, Veenhuizen AW, Zhang YY, Isakson PC (1997) J Med Chem 40:1347–1365
47. Weitzel KW, Wickman JM, Augustin SG, Strom JG (2000) Clin Ther 22:1254–1267
48. Zablocki J, Palle V, Blackburn B, Elzein E, Nudelman G, Gothe S, Gao Z, Li Z, Meyer S, Belardinelli L (2001) Nucleosides Nucleotides Nucleic Acids 20:343–360
49. Kısmet K, Akay MT, Abbasoğlu O, Ercan A (2004) Cancer Detect Prev 28:127–142
50. Cui JJ, Tran-Dubé M, Shen H, Nambu M, Kung P-P, Pairish M, Jia L, Meng J, Funk L, Botrous I, McTigue M, Grodsky N, Ryan K, Padrique E, Alton G, Timofeevski S, Yamazaki S, Li Q, Zou H, Christensen J, Mroczkowski B, Bender S, Kania RS, Edwards MP (2011) J Med Chem 54:6342–6363
51. Zora M, Kivrak A, Yazici C (2011) J Org Chem 76:6726–6742
52. Su M, Tomás-Gamasa M, Serdjukow S, Mayer P, Carell T (2014) Chem Commun 50:409–411

53. Biao Wu ST, Lin L (2016) vol CN 106432082 A
54. Cunqing Pan CQ, Shunlin Z, Yaoguo C (2017) vol CN 107098860 A
55. Ruibo Xu PZ, Xiaotian Y, Yiqiu Q (2018) vol CN 108840861 A
56. Ma J, Yang J, Yan K, Sun X, Wei W, Tian L, Wen J (2021) Eur J Org Chem 2021:5491–5496
57. Ma J, Yang J, Yan K, Luo B, Huang K, Wu Z, Zhou Y, Zhu S, Zhao X-E, Wen J (2023) SynOpen 07:272–276
58. Wang H, Gao X, Lv Z, Abdelilah T, Lei A (2019) Chem Rev 119:6769–6787
59. Wen J, Qin H, Yan K, Yang X, Sun X, Wei W, Yang J, Wang H (2020) Org Lett 22:8824–8828
60. O'Hagan D (2000) Nat Prod Rep 17:435–446
61. Harrison TS, Scott LJ (2005) Drugs 65:2309–2336
62. Venukrishnan Komanduri CDG, Krische MJ (2008) J Am Chem Soc 130:12592–12593
63. Bull JA, Mousseau JJ, Pelletier G, Charette AB (2012) Chem Rev 112:2642–2713
64. Vitaku E, Smith DT, Njardarson JT (2014) J Med Chem 57:10257–10274
65. Guan A-Y, Liu C-L, Sun X-F, Xie Y, Wang M-A (2016) Biorg Med Chem 24:342–353
66. Lee KN, Lei Z, Ngai M-Y (2017) J Am Chem Soc 139:5003–5006
67. Wen J, Yang X, Yan K, Qin H, Ma J, Sun X, Yang J, Wang H (2021) Org Lett 23:1081–1085
68. Proctor RSJ, Phipps RJ (2019) Angew Chem Int Ed 58:13666–13699
69. Niu C, Yang J, Yan K, Xie J, Jiang W, Li B, Wen J (2022) iScience 25:104253
70. Dunaev V, Aleksandrova E, Krasovskii A, Milonova N, Tishkin V, Linenko V (1986) Pharm Chem J 20:722–725
71. Krenitsky TA, Hall WW, Selph JL, Truax JF, Vinegar R (1989) J Med Chem 32:1471–1475
72. Smolii O, Gorodetskova N, Brovarets V, Klebanov B, Drach B (1989) Pharm Chem J 23:899–902
73. Lange JH, van der Neut MA, Borst AJ, Yildirim M, van Stuivenberg HH, van Vliet BJ, Kruse CG (2010) Bioorg Med Chem Lett 20:2770–2775
74. Zheng X, Ma Z, Zhang D (2020) Pharmaceuticals 13:37–56
75. Yang J, Zeng T, Yan K, Qin Z, Wen J (2022) Adv Synth Catal 364:1677–1682
76. Zeng T, Yang J, Yan K, Wei W, Wen J (2023) Green Synth Catal 4:35–40
77. Shi R, Lu L, Zhang H, Chen B, Sha Y, Liu C, Lei A (2013) Angew Chem Int Ed 52:10582–10585
78. Ouyang K, Hao W, Zhang W-X, Xi Z (2015) Chem Rev 115:12045–12090
79. Shi R, Zhang H, Lu L, Gan P, Sha Y, Zhang H, Liu Q, Beller M, Lei A (2015) Chem Commun 51:3247–3250
80. Wang Q, Su Y, Li L, Huang H (2016) Chem Soc Rev 45:1257–1272
81. Gao B, Huang H (2017) Org Lett 19:6260–6263
82. Shi R, Niu H, Lu L, Lei A (2017) Chem Commun 53:1908–1911
83. Zeng T, Yang J, Yan K, Wang S, Zhu S, Zhao X-E, Li D, Wen J (2022) Org Chem Front 9: 6305–6311
84. Murdock LL, Hopkins TL (1968) J Agric Food Chem 16:954–958
85. Hadaway AB, Barlow F, Turner CR, Flower LS (1977) Pestic Sci 8:172–176
86. Leader H, Casida JE (1982) J Agric Food Chem 30:546–551
87. Rodgers KE (2001) Krieger RI, Krieger WC (eds) Handbook of pesticide toxicology2nd edn. Academic Press, San Diego
88. Kaboudin B, Emadi S, Hadizadeh A (2009) Bioorg Chem 37:101–105
89. Riches J (2015) Timperley CM (ed) Best synthetic methods. Academic Press, Oxford
90. Turner JA (2015) The pesticide manual.17th edn. British Crop Protection Council, Hampshire, pp 1–1440
91. Farha W, Abd El-Aty AM, Rahman MM, Shin HC, Shim JH (2016) Environ Monit Assess 188:693–714
92. Nicolopoulou-Stamati P, Maipas S, Kotampasi C, Stamatis P, Hens L (2016) Front Public Health 4:148–156
93. Li X, Yang X, Zheng X, Bai M, Hu D (2020) Int J Mol Sci 21:7144–7160

94. Yin Y, Yang J, Yan K, Zeng T, Lin H, Ling J, Wang S, Wen J (2024) Green Chem. https://doi.org/10.1039/D1033GC02727A
95. Pohjala L, Alakurtti S, Ahola T, Yli-Kauhaluoma J, Tammela P (2009) J Nat Prod 72:1917–1926
96. Cho W, Koo JY, Park Y, Oh K, Lee S, Song J-S, Bae MA, Lim D, Lee D-S, Park SB (2017) J Med Chem 60:170–179
97. Chen Y-J, Zhang G-Y, He Y-H, Guan Z (2019) Cat Sci Technol 9:4216–4221
98. Yang J, Song M, Zhou H, Qi Y, Ma B, Wang X-C (2021) Green Chem 23:5806–5811
99. Li C, Zhong Q, Tang S, Wang L, Li P, Li H (2022) Org Chem Front 9:3769–3774
100. Wen J, Ma J, Yan K, Zhang X, Huang K, Wu L, Li D, Yang J (2023) ACS Sustain Chem Eng 11:16785–16792
101. Guang H, Ji-Chun Y, Hui-Chao LI, Jing Z, Chang-Ling L (2011) Agrochemicals 50:79–82
102. Liu A, Wang X, Liu X, Li J, Chen H, Hu L, Yu W, He L, Liu W, Huang M (2017) J Heterocyclic Chem 54:1625–1629
103. Li D, Chen L, Jin Y, Wang X, Liu L, Li Y, Chen G, Wu G, Qin Y, Yang L, Wang M, Zhao L, Xu Z, Wen J (2023) Green Chem 25:4656–4661

SPRINGER NATURE

GPSR Compliance

The European Union's (EU) General Product Safety Regulation (GPSR) is a set of rules that requires consumer products to be safe and our obligations to ensure this.

If you have any concerns about our products, you can contact us on ProductSafety@springernature.com

In case Publisher is established outside the EU, the EU authorized representative is:

Springer Nature Customer Service Center GmbH
Europaplatz 3
69115 Heidelberg, Germany

The manufacturer's authorised representative in the EU is Springer Nature Customer Service Centre GmbH, Europaplatz 3, 69115 Heidelberg, Germany. If you have any concerns regarding our products, please contact ProductSafety@springernature.com

Printed and bound by CPI Group (UK) Ltd, Croydon, CR0 4YY
06/10/2026
02225878-0005